Glasses for Infrared Optics

Valentina F. Kokorina

CRC Press
Boca Raton   New York   London   Tokyo

The CRC Press
Laser and Optical Science and Technology Series

Editor-in-Chief: Marvin J. Weber

Handbooks of Laser Science and Technology
Volume I: Lasers and Masers, Edited by Marvin J. Weber
Volume II: Gas Lasers, Edited by Marvin J. Weber
Volume III: Optical Materials, Part 1, Edited by Marvin J. Weber
Volume IV: Optical Materials, Part 2, Edited by Marvin J. Weber
Volume V: Optical Materials, Part 3, Edited by Marvin J. Weber
Supplement I: Lasers, Edited by Marvin J. Weber
Supplement II: Optical Materials, Edited by Marvin J. Weber

Crystalline Lasers:
Physical Processes and Operating Schemes
Alexander A. Kaminskii

Imaging Through Turbulence
Michael Roggemann and Byron Welsh

Optical Constants of Inorganic Glasses
Andrei M. Efimov

Spontaneous Emission and Laser Oscillation
in Microcavities
Hiroyuki Yokoyama and Kikuo Ujihara

Thermodynamic and Kinetic Aspects
of the Vitreous State
Sergei V. Nemilov

Forthcoming Handbook Titles
Handbook of Laser Wavelengths, Edited by Marvin J. Weber

Glasses for Infrared Optics

Aquiring Editor:	Tim Pletscher
Project Editor:	Jennifer Richardson
Marketing Manager:	Susie Carlisle
Direct Marketing Manager:	Becky McEldowney
Cover Designer:	Dawn Boyd
Prepress:	Kevin Luong
Manufacturing:	Sheri Schwartz

Library of Congress Cataloging-in-Publication Data

Kokorina, Valentina F.
    Glasses for infrared optics   /   by Valentina F. Kokorina.
       p.   cm. -- (Laser and optical science and technology series)
    Includes bibliographical references and index.
    ISBN 0-8493-3785-2 (alk. paper)
    1. Infrared detectors--Materials.   2. Optical materials.
   3. Glass.   4. Chalcogenides.   I. Title.   II. Series: CRC Press
laser and optical science and technology series.
TA1570.K65   1996
621.36'2--dc20

    95-39604
    CIP

No claim to original U.S. Government works
International Standard Book Number 0-8493-3785-2
Library of Congress Card Number 95-39604
Printed in the United States of America 1 2 3 4 5 6 7 8 9 0
Printed on acid-free paper

The Author

Valentina Feodorovna Kokorina was born in Russia. In 1944 she entered the Leningrad Technological Institute and in 1949 received an M.A. diploma of the highest degree. Her specialization was chemistry and technology of glass. She then began work in the laboratory for physicochemical properties of glass in the S. I. Vavilov State Optical Institute in Leningrad. Soon thereafter she began post-graduate studies and received a Ph.D. degree in chemical sciences in 1954. Since 1955 she has been studying the structure and properties of oxygen-free chalcogenide glasses. In the 1960s she was head of a laboratory and group of scientists working with chalcogenide glasses. Their investigations resulted in the development of various optical materials for use in the infrared spectral region. In 1969 she became a Doctor of Technical Science; her dissertation was entitled "Properties, Structure, and Manufacturing of Optical Chalcogenide Glasses". From 1970 to 1974 Dr. Kokorina was vice-director of science in one of the divisions of the State Optical Institute (now the Scientific-Research Technological Institute of Optical Materials Science). During the following 12 years she headed the laboratory for special glasses and glass-ceramics in the Institute. In 1980 she and four other members of her group received the State Prize for the development and mass production of commercial chalcogenide glasses.

Dr. Kokorina has received author licenses for 18 inventions, the majority of which are related to manufacturing. Based on these and her other activities, she was awarded the degree of honored inventor of the USSR. Dr. Kokorina is the author of over 60 publications and one monograph and has presented reports at many scientific conferences in Russia and abroad. Seven dissertations on chalcogenide glasses have been completed under her supervision.

Contents

Chapter 2
Optical Properties of Chalcogenide Glasses

Chapter 3
Elaboration of Commercial Glasses

Chapter 4

Technological Basics for Manufacturing Optical Chalcogenide Glasses

Preface

Infrared optics are useful for many potential applications such as diagnostics in medicine, nondestructive testing in industry, thermal scanning of the ground for ecology aims, and everywhere when identification of objects according to their surface heating patterns is necessary. In all of these applications infrared (IR) optics are in need of high-quality optic materials for its further development. Of most importance is the transparency of materials in 2- to 14-μm spectral range, for in this wavelength range lies the major part of energy of radiation emitted by objects at temperatures of practical interest. Here are the so-called "atmospheric windows of transparency", determined by the position of absorption bands of molecules of water (vapor) and carbon dioxide contained in the atmosphere.

Nowadays infrared transparent crystals are usually used in thermovision systems. Crystalline media can also be used in optical instruments for the visible spectrum range. Although many of such crystals are known, optical glass appears to be the principal material used in the majority of instruments. In fact, glasses have certain advantages over crystals as the material for optical systems. The technology of their manufacturing is comparatively easy and more available for mass production. It secures lower cost of goods and assumes large-size detail production; and above all, glasses provide a broad variety of compositions and consequently of optical properties, which allows one to create more perfect instruments.

Unfortunately, the most widely distributed glasses — silicates, borates, and phosphates — are transparent only in the near-IR range, i.e., up to 3 μm, and highly pure quartz glass, up to 4 μm. Glasses based on germanium dioxide which transmit radiation of the middle-IR range (up to 6 μm) are well known. This is the transparency limit for oxygen-containing glasses, because the vibration frequency of oxygen bonds with any other element of the periodic system approximates these very wavelengths. The frequency of the simple harmonic oscillation is expressed by the formula:

$$\nu = 1/2\,\pi C \sqrt{K/M}$$

where M is effective mass of oscillating nuclei. Hence, vibrational absorption bands of substances must shift to long waves with the increase of atomic weight of constituent elements. Therefore the transparency edge of glasses can be shifted to the long-wave spectrum range through oxygen replacement with the elements of the seventh group, halogens, or with its analogues in the subgroup, chalcogens. In recent years extensive investigation has been carried out of the glasses based on fluorine compounds with heavy metals. However, their rather high crystallization tendency poses certain problems for industrial manufacturing, insuperable at the present time. In this way, oxygen-free glasses based on sulfur, selenium, and tellurium still represent the only commercial vitreous material transparent in the long-wave IR range. They are called by convention "chalcogenide glasses" (ChG) — by analogy with oxide ones. Although this term doesn't embrace all opportunities of glass formation in this class of substances, it is generally accepted nowadays and we will use it in this book. ChG are widely known as vitreous semiconductors and also as the substances which manifest photo-inducted changes of properties. In this term their global investigation began after 1955, when Goryunova and Kolomiiets discovered electronic conductivity in vitreous substances obtained from mixtures of arsenic, antimony, and thallium chalcogenides at the synthesis of crystal semiconductors.[1,25] These substances got the name of chalcogenide vitreous semiconductors. Many articles and several monographs are dedicated to the physics and chemistry of chalcogenide vitreous semiconductors. The results of the investigations of their structure, electrical, and other physical properties are presented in the monograph by Mott and Davis.[2] The same problems are reviewed in the collective monograph edited by Brodsky.[3] Chemistry of Semiconductor Glasses, according to the evidence reported from the investigations carried out in the Leningrad State University, is observed in two books by Borisova.[4,5] An earlier collection of research also exists carried out in the Leningrad University under the guidance of Myuller.[6] Investigations of the chemistry of ChG are being carried out now in the Institute of General Inorganic Chemistry named after Kurnakov. Thus, in 1984 the monograph by Vinogradova was published where the author presents information about glass forming and phase equilibrium in 25 binary and more than 150 ternary chalcogenide systems.[7] This list is far from being complete, but it shows quite clearly that research of chalcogenide glasses is widely presented in the literature.

The ways for obtaining such substances are but cursorily touched in the above-mentioned publications. More attention (which, however, in our opinion, still isn't enough) is paid to this problem in the monograph by Feltz, where chalcogenide glasses are examined along with amorphous and vitreous solid matters,[8] and also in the book by Minayev, published in 1991, where the evidence from the investigations of almost all known vitreous semiconductor melts is reported.[9] These monographs also present a thorough literary review. However, no monographs are known where chalcogenide glasses are treated as an extensive independent class of vitreous substances with distinctive physicochemical and optical properties and where the problems of their synthesis technology are discussed.

Meanwhile, the technological process of glass manufacturing is the most important aspect of the problem of vitreous substance application. An incorrectly elaborated technological process — its slightest violation — leads to nonreproductivity of composition, hence properties of glasses from different batches, i.e., makes their mass application impossible. It is particularly important for ChG, for their specific character doesn't allow one to manufacture them in large amounts simultaneously. That is why the problems of their technology, including the manufacturing of glasses with easily producible compositions, are discussed in detail in this book and this is the first attempt of its kind.

The material for this book was taken from research works which, from 1955 up to the present, have been carried out in the State Optical Institute named after S.I. Vavilov: at first under the guidance of K.S. Yevstropiev and R.I. Myuller, and later under the guidance of the author. It is necessary to say that the major part of these works, dedicated to the synthesis and investigation of ChG for the manufacturing of commercial materials for IR optics, had been done in the 1960s to 1980s, but, judging from the literary data, they still present novelty. Unfortunately, for some reason or other, many of the results couldn't be published. The present book aims to meet this lack. According to the stated problem the primary attention in all chapters of the book is paid to the author's research and the period of time when it was carried out is mentioned when necessary. The literary material is observed only in historical aspect for the purposes of comparative analysis of the results. The references to the author's own publications are given in the beginning of each chapter. This list is placed after the bibliography.

I consider it necessary to mention separately those research workers who, together with me, traveled a long way in the creation of the new optical material — chalcogenide glasses. This way began with the extraction of little pieces weighing 5 to 10 g and ended with large-tonnage manufacturing. The research and work of these people determined to a large extent the contents of this book. They are L.G. Aijo, Ye. A. Kicutzkaya, A.M. Yefimov, and V.V. Melnikov. In 1980 they all became winners of the U.S.S.R. State Prize. Those who will be named our successors have already done a lot of work and have a lot to do in the future. They are A.V. Belykh, M.D. Michailov (head of a group), who, one way or another, helped me in writing this book — and I'm very obliged to them for that. I'm also very grateful to N.N. Zandina who has found and translated literary original sources. Special thanks to A. Kokorina, my granddaughter who took the trouble of translating this book into English.

Introduction

Mankind has been using glass for more than 4000 years and all this time the vitreous state was considered to be the prerogative of oxygen compounds and their blends. It was only at the end of the 19th century when vitreous selenium and also arsenic sulfide and selenide were obtained for the first time, but even then they didn't attract investigators' attention. Active research of new compositions and ways of synthesis was initially connected with their peculiar optical properties, namely, transparency for middle and far infrared (IR) radiation, as the impetuous development of IR optics made it necessary to produce new IR materials. That is why from 1950 to 1953 a series of research works appeared on the synthesis and exploration of optical properties of amorphous selenium, arsenic sulfide, and arsenic selenide — both pure and with additions of other chalcogenides. We note the research works by Frerichs,[10,11] Fraser,[12] and Dewulf,[13] who put these substances forward as IR-optical materials. In 1957 in the U.S. Glaze and others worked out half-commercial technology of the vitreous As_2S_3 manufacturing, and the "Servo Corporation" company began to advertise it as a commercial material "Servofrax".[14]

Later pure selenium and selenium with 10% arsenic impurity were also proposed as industrial materials, but these systems are of little use due to their poor hardness and low softening temperature. Intensive investigations of chalcogenide glass (ChG) were continuing, and it should be noted that up to 1958 in all the research only chalcogenide mixtures were used — by analogy with oxide glasses as oxide mixtures. Even with the application of elementary sulfur, selenium, and arsenics as initial materials, the compositions were calculated in stoichiometric ratio.

Meanwhile, due to the presence of such glass-formers as sulfur and selenium and also blends of tellurium with selenium, glasses are formed both in the region with chalcogen excess against stoichiometry and in the region with its lack, which is absolutely impossible with oxide glasses. Hence it would be more correct to name these substances "oxygen-free glasses based on chalcogens".

In the State Optical Institute named after S.I. Vavilov the investigation of chalcogenide glasses began in 1955 with the synthesis of As_2S_3 and As_2S_3 with Sb_2S_3 impurities. It appeared to be possible to obtain some pseudobinary compositions in vitreous state and to study their properties, including transparency.

However, already in 1958 the idea of using selenium as a glass-former was realized for the first time and it enabled us to synthesize and study glasses in the systems consisting of only two elements, namely, As–Se, P–Se, Ge–Se, etc., and to produce multicomponent glasses on their basis — both with the excess and lack of chalcogen. As a result we studied the glass-forming tendency and physicochemical properties of ChG of more than 40 elementary and complex systems, including 27 chemical elements. A detailed description of the laboratory methods for the synthesis of glasses and the ways of their homogeneity control, elaborated for these purposes, is given in Chapter 1. The methods for conducting the measurements of properties used in our works — their precision and reproducibility — are also described in this chapter. The results of the measurements are presented in the tables and on the diagrams "composition-property".

Peculiar properties of ChG and their dependency on composition are determined by the structural peculiarities of these vitreous substances. Thus, unlike the majority of oxide glasses, chalcogenide ones don't include, as a rule, the ionic component of chemical bond. The possibility of studying systems which consist of only two elements with the chemical bond of one type only advanced us considerably in our idea of the nature of glass-forming and glass structure. The author has to confess that the structural properties of substances necessary for glass-forming were described through empirical methods. However, as is generally known, hitherto there are no universal quantitative criteria that could allow one to calculate coordination atomic numbers necessary for glass-forming, describe population of chemical bonds in glasses, and connect it with their real properties.

Pointing out major achievements in investigations of the structure of amorphous and vitreous substances by physical methods, Feltz writes, in particular: "We are dealing in the main with the qualitative rules. It must be noted that on the presently achieved level of theoretical development it is still necessary to be careful with the results of quantitative calculations even in the cases when they are in satisfactory concord with an experiment."[8]

In the 1970s to 1980s several works appeared in which the authors strived to put forward quantitative criteria of glass-forming. Thus, Stanworth relates glass-forming to the degree of spatial filing (f), which is calculated from crystal density, molecular mass, and orbital atomic radii of glass components.[15] The author has demonstrated that in glass-forming oxides AxOy the criterion f = 0.13 to 0.24.

This criterion, however, is worse adopted to ChG. Thus, it doesn't mark the difference of glass-forming tendency in the row As_2S_3, As_2Se_3, As_2Te_3 (f, respectively, 0.15, 0.14, 0.20). Meanwhile this difference, first established by Kolomiiets and Gozyunova,[16] undoubtedly exists and it is the special case of decrease of glass-forming with increase of elements' atomic number.

Well-known is Phillips' topological criterion based on calculation of sums of tensile and transverse stresses of covalent bonds, falling on one atom (N_c). According to Philips,[17] the maximum mechanical stability in coordinate space is achieved at $N_c = 2.45$ and that appropriates to the maximum glass-forming ability. Philips' criterion enables us to separate the nonglass-forming chalcogenides As_2Te_3 and $GeSe_2$ from the glass-forming ones. However, selenium, which is known as a perfect glass-former, doesn't fit this criterion.

Other criteria — by Dietzel[18] and Baydakov and Blinov,[19] aren't universal as well and can't be adapted to multicomponent glasses.

There exist at least seven known qualitative kinetic criteria; it isn't necessary to analyze them here because it has been done in detail in the monograph by Dembovsky and Chechetkina,[20] where the glass-forming problem is examined in structure-chemical, thermodynamical, and kinetic aspects. The two last ones, however, concern essentially not glass-forming, but glass transition, i.e., the process of "crystal-glass" transition, and deal with energy difference between vitreous and crystalline states. The authors operate with glass-transition temperature, critical rate of cooling, energy barriers of crystals' origin and growth, etc. However, these are nothing but substance properties and as the glass-forming tendency is also a property of the given substance, all the kinetic criteria come to the explanation of one property change with the help of another, and it doesn't do a lot for the solution of the problem of glass-forming. Thus, for example, answering the question of why SiO_2 easily transforms into glass and SiS_2 doesn't, we can say that they differ in crystallization rate or in viscosity at softening temperature. However, the question of why these properties of the given substances are also so different remains open.

It is quite obvious that for the successful approach to cognition of the nature of the vitreous state and glass structure we must relate thermodynamics and kinetics of glass transition to the structure-chemical peculiarities of glass-forming substances. Anyway, while making new glasses with certain properties it's necessary to have at least an empirical idea, not only about coordinating numbers and the character of component bonds of the researched system, but also about the changes of these parameters due to the reactions going in the process of synthesis, i.e., about the glass structure.

The investigation of the dependence of various properties on composition, the analysis of glass-forming region, and constitutional diagrams can present reliable information about the basic features of glass structure and its changes with composition. As to direct establishment of structure with the help of diffraction methods, although comparatively progressive, when it concerns glasses it meets serious difficulties due to the high erosion of the diffraction pattern. Application of methods of IR spectroscopy is limited by the shortage of absorption bands in the spectra of glasses, which also, as a rule, are highly diffused. That's why a number of questions about the structure of even so thoroughly studied glasses as silicate and borate ones remain open. That concerns, in particular, dimensions and role of the order region. Under such conditions the information about ChG structure that can be gotten from the analysis of "composition-property" diagrams is of great value. It's necessary to note the following here. It is generally accepted to divide the properties of glasses into structure-sensitive and structure-insensitive ones. It seems to be unwarranted that any property, low-susceptible to structural change within one class of glasses, appears to be structure-sensitive as applied to the glasses of another class. Besides, it is inconceivable that the most abrupt structural transformations taking place during chemical reactions in the process of glass-forming and connected with the change of near order could not influence such properties of glasses as density, viscosity, coefficient of thermal expansion, etc. One can see from the experimental data in Chapter 1 that on the diagrams which show how the properties change with composition, both chalcogenide and oxide glasses have characteristic bends and fractures, coming not only from the presence of chemical compounds and eutectics.

The analysis has shown that structural changes deriving from emergence and quantitative accumulation of other structural units are marked on the curves of properties with peculiar peculiar points. We were also first to establish that the magnitudes of glass properties with components of different nature and composition coincide if the glasses have identical covalent or energy atomic coordination.

Based on the whole complex of the investigative results, the structural model of vitreous substances was put forward and structural peculiarities of the initiate substances, necessary for glass-forming, were formulated. Therewith the glass-forming tendency of substances was considered as the positional function of the compound elements in the periodic system.

Chapter 2 is dedicated to the optical properties of ChG. As was mentioned above, the main distinction of these glasses, which determines their value as optical materials, is their high transparency for IR radiation in the wide wavelength intervals: from the short-range infrared to the radio range (Figure 1).

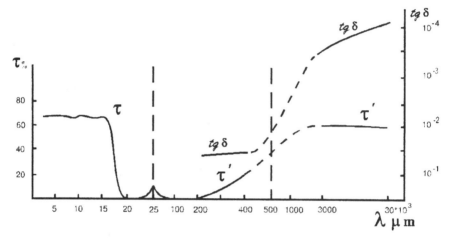

FIGURE 1. The general view of the transmission spectrum of chalcogenide glasses given by the example of $As_{30}Ge_{25}Se_{45}$. Thickness 1 mm.

Up to 1960 only arsenic chalcogenides and selenium were the objects for investigation of IR spectra of ChG. As a result spectral characteristics described in different works performed a rather mixed view when concerned with both the number of absorption bands and their intensity. Spectral transmission of glasses, containing chalcogenides of arsenic, antimony, and bismuth, had been investigated for the first time by Kolomiiets and Pavlov.[21] The authors demonstrated that in the series there takes place consecutive transmission shift into the long-wave spectrum range. However, in this work also observed were the mixed character of transmission magnitude and position of absorption bands. Later the explorers set themselves the task of dividing bands into intrinsic (deriving from oscillations of atoms of glass components) and extrinsic ones, and also of identification of the extrinsic bands.

In Chapter 2 an account is given referring to the 1963 to 1964 research work by A.M. Yefimov, who examined in detail intrinsic absorption bands and optical losses — both due to extrinsic atoms and molecules and to the presence of the second phase in the glass volume. Also cited there are the results of the analysis of the literary data concerning identification of absorption bands, obtained lately.

Besides spectral transparency, the optical properties of glasses also include the refractive index and its dependence on radiation wavelength, i.e., on dispersion. As the refractive index changes incessantly with the wavelength, to evaluate the difference in dispersion of various glasses, Abbe introduced the notions "middle dispersion" and "dispersion coefficient" also known as the Abbe number.

To characterize glasses in visible spectrum range the so-called middle or principal dispersion $\Delta n = n_{F'} - n_{C'}$ is used, where F' and C' are spectral lines of cadmium (479.99 and 643.85 nm, respectively). The calculation of the dispersion coefficient is performed by the formula $v_e = (n_e - 1)/(n_{F'} - n_{C'})$, where n_e is the refractive index at the wavelength in mercury spectrum (546.07 nm); n_e and v_e are optical constants; calculations of the optical systems are performed by them.

The majority of ChG is absolutely or almost opaque in the visible spectrum range, hence conventional constants, different in different works, are used for them.

In Chapter 2 results, first performed by us, of the systematical investigation of the changes of optical constants with composition of glasses of elementary and complex systems based on chalcogens are described. The work that had been done made it possible to calculate changes of optical constants introduced by various elements, taking into account their permissible content in ChG, and to plot the Abbe diagram in coordinates $n_{2.0} - v_{2.0}$.

In Chapter 3 results are presented of the investigations connected with manufacturing of commercial ChG. It concerns glasses used for production of lens, prisms, optical wedges, protective plates, and semispheres — transparent in different wavelength ranges (0.7 to 9, 0.8 to 11, 1 to 15.5, and 1.5 to 17 μm); pairs of glasses for objectives of optical systems; glasses for passive elements of IR lasers; glasses for fiber optics; IR filters of various types. Of special interest is the section dedicated to the elaboration of fusible commercial glasses containing the elements of the VII group of the periodic system.

The peculiarities of industrial technology of chalcogenide optical glasses in comparison to oxide ones are presented in Chapter 4.

The technological process that we had elaborated together with technologists of LZOS allows us to produce castings weighing up to 24 kg. On LZOS the glasses in blanks of various size and shape are manufactured. At maximum the cylindrical blanks are 350 mm across, 35 mm thick. Instructions for safety measures during manufacturing and cool working of ChG are added to the technological process.

It is necessary to emphasize here a very important aspect of glass applications. The analysis of the literary data has formed an impression that ChG have been lately used in IR optics in lesser extent than they could be. If our readers look through this book attentively, they'll understand perfectly well that such a situation doesn't help quality improvement and reduction of prices for IR devices, hence it doesn't help further development of the branch.

Designers and manufacturers of the devices are on the alert with these comparatively new materials, mainly owing to the legend about elevated toxicity of ChG caused by the presence of arsenic in their composition. However, this legend, like any other one, derives from lack of information. In fact, elementary arsenic is as toxic as its analogues in the periodic system — phosphorus, antimony, and bismuth — and not more toxic than other elements widely used in chemistry, in particular, in glass manufacturing. Virulent poison is not arsenic, but it's oxygen compound As_2O_3 — white powder which dissolves easily in water and acids. However, even traces of oxygen are intolerable in ChG; this is obtained by the vacuum method of synthesis and special purification of raw materials. In ChG arsenic combines with sulfur, selenium, tellurium, or with metals that enter into the composition of glasses. These arsenic compounds don't dissolve either in water or in acids, hence they are not toxic. It is well known that arsenic sulfide As_2S_3 is used as a dye, hence its name — oripigment or yellow arsenic. As_2S_3 and As_2Te_3 also don't refer to poisons. Glasses containing thallium may be slightly toxic; that's why there is no thallium in our commercial glasses. Workers worry about the strong smell during cool treatment of ChG, but it has nothing to do with arsenic. This smell appears as a result of the reaction of sulfur and selenium with water in the process of grinding, when hydrogen sulfide and/or hydrogen selenide escape into the air. Appropriate safety measures help to reduce the concentration of these bad smelling gases into the atmosphere to the maximum permissible. The application of these glasses isn't more dangerous than application of any other IR materials, in particular, ZnS, ZnSe, CdTe, etc.

In Chapter 4 are described the properties and normalized quality parameters of IR glasses, included in the 1990 catalog, and also of new glasses for which technical documentation is not yet available. Also included is the comparative analysis of all presently known commercial glasses. ChG, manufactured by different companies, and optical constants of these are presented as a diagram $n_{10.0} - v_{10.0}$.

In conclusion, a short review of some trends in application of ChG as materials for IR optics is given.

1 Chalcogenide Glasses — A Peculiar Class of Vitreous Substances

1.1 METHODS FOR SYNTHESIS OF CHALCOGENIDE GLASSES[1]

ChG, with high transparency and stable physicochemical properties, must not contain oxygen impurities. Therefore, all known methods for synthesis of ChG can be combined on the basis of the two following principal indications:

1. Synthesis in the flow of inert gases or other oxygen-free gaseous substances
2. Synthesis in evacuated vessels in the atmosphere of inherent vapors

The first group represents the distillation of As_2S_3 in nitrogen flow with mixing by a quartz mixer in the casting chamber, offered by Glaze and others.[14] The authors obtained up to 5 kg of glass following this technological process. Frerichs also produced glass through distillation of crystal orpiment,[10] but in melting atmosphere he used hydrogen sulfide. To reduce blisters in As_2S_3 after distillation, Matsuda carried out the repeated glass heating in a quartz vessel with an opening in the bottom for a casting, with glass melt stirring by a quartz mixer at a speed of 10 rpm for 2 h.[22]

Deeg applied similar methods,[23] but instead of a crucible he used a quartz flask with a ground-in plug, with one opening for a mixer and two more openings for gas inlet or vacuum creation. Pearson and others, who were the first to study ternary systems with nonstoichiometric element relations, propose in their work to synthesize glasses in nitrogen or argon atmosphere in crucibles with lids, mixing glass melt in the process.[24] This method allows the producing of glasses of various compounds, but introduction of infusible elements is limited due to selective volatility of charge components — mainly because of chalcogens. The drawback, common for this group of methods, is the atmosphere pollution with arsenic, sulfur, and selenium vapors which escapes from the melting vessels together with exhausted inert gas. Arsenic oxide is formed easily in these conditions and it makes the methods described above especially complicated in respect to safety measures.

The second group of methods, which stipulates oxygen driven out by evacuation of melting vessels with charge, is free of this drawback.

In 1955 the State Optical Institute, named after S.I. Vavilov, L. Aijo, and
V. Slavyanski, created the plant for orpiment distillation under vacuum. The process
of synthesis was carried out in specially shaped Pyrex® ampules which allowed the
homogenizing of glass melt by rotation of an ampul about the longitudinal axis.
Although this method of sublimation is preferable when a gas atmosphere is used,
it also doesn't allow the producing of glasses of various complex compositions.
Dewulf was the first who used the synthesis of ChG in evacuated vessels of Pyrex®
glass and by this method she had produced the glasses from mixtures of sulfides
and selenides of the elements of the V group.[13] A similar method was used by
Goryunova and Kolomiiets in the synthesis of crystal semiconductors and later of
ChG.[25,26] They filled a quartz ampul with the charge and hooked it up to the vacuum
system. An evacuated and unsoldered ampul was put in vertical position into the
electric furnace and after being held there under the required conditions it was
exposed to the air. Melting in the evacuated vessel allowed the introduction of various
elements in any proportions into the charge and therefore the producing of glasses
of various compositions. It ruled out the possibility of the components' volatility,
hence the true composition appeared to be identical to the calculated one. However,
it was impossible to synthesize germanium glasses, for they segregated to a certain
extent. Besides, it is known that homogeneity of glasses, necessary for their application in optics, can be achieved only during the homogenizing of glass melt in the
process of synthesis. As was demonstrated above, this mode had been used by
practically all investigators who synthesized ChG in the flow of inert gases. However,
glass melt, alloying in the process of synthesis in evacuated vessels, proved to be
rather complicated. Nevertheless, in 1960 we had already worked out the laboratory
method for the melting and manufacturing of ChG which allowed the production of
rather large-sized samples, homogeneous with a minimum number of defects and
an extended composition range. Because this technique has formed the basis of the
synthesis methods for such glass applications in the entire world since 1962, we
consider it necessary to reproduce it in detail.

1.1.1 Laboratory Method for the Synthesis of Optical Chalcogenide Glasses

Initial materials (mainly the elements of the IV, V, VI, and VII groups of the periodic
system) must contain not more than 0.01 at.% of impurities. Otherwise they should
be exposed to a supplementary purification procedure. The charge is prepared by
grinding the components in porcelain or metal stamps (or by other means). Ground
substances are screened through 20 to 25 sieves, and after that the specimens accurate
to 10 mg, according to the glass percent composition, are taken out. Thoroughly
mixed charge is put into quartz ampules and hooked up to the vacuum system. The
dimensions of the ampules are determined by the quantity of the synthesized glass
with the allowance for overall melting furnace dimensions. In our laboratory furnaces
up to 500 g of glass can be melted. Prior to putting the charge on the intrinsic surfaces
of the ampules are thoroughly cleaned by sequential washing by hydrofluoric acid,
water, and alcohol, connected to the vacuum system, and dried by warming up at 800
to 850°C in the vertical electric furnace in the process of persistent vacuum treatment.

The charge is put into the prepared vessels and after that it is again connected to the vacuum system and evacuated up to not less than 10^{-3} mm mercury. In the process of vacuum treatment the ampules with charge are warmed up at 250 to 350°C for several hours, depending on glass composition, for removal of moisture and gases adsorbed by charge materials. After the necessary vacuum is established, the ampules are unsoldered from the system with the help of an oxygen-hydrogen burner or a voltaic arc. The melting process is carried out in special rotating electric furnaces with thermoregulation-performing temperatures up to 1200°C. In the operative space of the furnace, gradient temperature should not exceed 10°C. For the maximum temperature leveling, the vessels with charge compressed by asbestos are put into the quartz pipe which is then placed horizontally in the furnace, for that provides a better compounding process due to the larger surface, and the depth of a glass melt layer is smaller.

The time-temperature regime of melting is worked out for each glass, respectively. The lowest temperature which is attained in the furnace in 30 to 60 min is, as a rule, 540 to 700°C. The melt is held at this temperature for 3 to 10 h, depending on charge amount, until glass-forming reactions come to the end. The clarification, or the upper melting temperature for the glasses free of the elements of the fourth group, is 800 to 900°C; the clarification period is 1 to 3 h. In the middle and at the end of the charge penetration process, and also at clarification temperature, glass melt homogenizing is carried out by rotating the furnace about the minor axis. After the last mixing the furnace is placed vertically and switched off. At the melting of glasses containing only arsenic chalcogenides, cooling is carried out spontaneously. For easily crystallizing glasses the necessary melt quenching is in the interval 450 to 500°C down to room temperature. For the synthesis of glasses containing the elements of the fourth group and of supplementary subgroups, the melting regime is to be slightly changed: the ampul with the charge is heated up to 600 to 700°C and held for 4 to 6 h. The clarification temperature for different glasses is 900 to 1100°C, held for 1 to 3 h. After that the ampules are cooled in the furnace down to 500 to 600°C and the glass melt is quenched.

Subsequent operations are identical for all glasses: after cooling down to the room temperature the melting ampul is sealed off and the ingot is visually inspected. If no visible crystallization is observed, the glass thermal treatment is to be held again for stress relieving and sample manufacturing. In laboratory conditions samples are, as a rule, prepared in the process of molling, when the glass is shaped by heating under gravity. This process is carried out as follows. Glass ingots are put into metal or quartz molds and heated in the muffle furnace. Molling also requires different time-temperature regimes for glasses of different compositions.

Glasses with low crystallizability are heated to a temperature slightly higher than the softening one. After the glass fills the mold uniformly, the muffle is switched on the annealing regime.

Glasses with high crystallizability are heated up to a temperature higher than that of liquids (melting temperature for all crystal phases of this glass), quickly cooled down to annealing temperature, and annealed. The annealing temperature is established for each group of glasses, respectively, according to dilatometric tests. The upper annealing edge is 10 to 15°C lower than the temperature of glass dilatometric softening. At the annealing temperature the glass is held for several hours (time interval is determined by the glass amount) and after that the temperature reduced at 10 to 20°C/h. The annealed samples are put through the cool treatment. The process of the cool treatment of ChG surfaces is similar to the one accepted for oxide glasses. However, some precautions are necessary, as the coefficient of thermal expansion is comparatively high and the hardness is lower than in oxide glasses.

Grinding is performed only with emery water suspension (15, 60, and 120 min). Final treatment by small emeries is performed on glass. Polishing is performed on velvet using chromium oxide water suspension. When a good plane is needed, polishing may be performed on tar.

The technique described above allows the producing of such amounts of glasses of such quality that it is quite possible to define glass-forming regions and to study their properties, and also to produce experimental batches for usual application. Changes and additions, necessary at the synthesis of especially pure glasses for laser and fiber optics, are exhibited in the applicable parts of Chapter 3.

1.1.2 SAFETY MEASURES FOR MANUFACTURING AND TREATMENT OF CHALCOGENIDE GLASSES

ChG involve chalcogens and also chalcogenides of the elements of the IV and V groups which dissolve at room temperature neither in water nor in acids. Hence even arsenic chalcogenides (unlike its soluble compounds) are not poisonous. Thus, arsenic ChG can be used as details of optical systems equally with oxide glasses without any particular safety measures. However, during glass synthesis and treatment, in order to avoid workers' poisoning with vapors and dust of the forming toxic substances, the following safety measures are necessary:

1. All stages of glass manufacturing — from the purification of raw materials and charge preparation to cool treatment of samples — must be performed in ventilation stoves or under hoods.
2. While working at furnaces and plants it is necessary to wear overalls and gloves; at a charge preparation one must put on a respirator.
3. Gloves must be on during charging vessels and when removing the produced glass.

In the process of glass melting in evacuated vessels, breakage of vessels in the furnace is possible, owing to the defects of the vessels or to the use of an improper temperature regime. Therefore, as a result of contact at the charge components with air oxygen, a certain amount of arsenic oxide and other toxic substances may escape into the atmosphere. To avoid that the following safety measures are necessary:

1. Examine the melting ampules thoroughly before charging (check the walls for thickness, different thickness, presence of scratches, and cuts). Check the vessels thoroughly after the sealing off from the melting plant, allowing no hollows, cavities, or reduction of wall thickness in the place of sealing off. Putting the vessels with defects into the furnace is forbidden!

2. Fix the melting ampules in the furnace properly in order to avoid their shift at mixing.
3. Carefully track the work of thermocouples and master apparatuses, preventing the furnaces from overheating.
4. Melt glasses only with the ventilation switched on and the stove doors closed.
5. In case of the furnace smudging, turn the electric current off, don the mask, and take measures to cool the melting space as soon as possible.
6. After each case of breakage of melting ampules, clean the walls of the vent hood thoroughly and change gas filters.

At the cool treatment of ChG gaseous products of the reaction of glass components with water — hydrogen sulfide and hydrogen selenide — can appear, which are toxic when their concentration in the air is over 0.01 ml/l.

To avoid air pollution exceeding the prescribed norms, ventilation from every bench in the area of the cool treatment is necessary. Further, the amount of toxic substances can be reduced by the addition of hydrogen peroxide to the polishing suspense ($H_2S + H_2O_2\ H \to 2\ H_2O + S$).

These recommendations form the basis of the Safety Measures at ChG Cool Treatment Instruction, which is presented in Chapter 4.

1.2 EVALUATION OF THE QUALITY OF CHALCOGENIDE GLASSES[2]

The absence of the crystalline phase and other defects in oxide glasses is easily seen with the naked eye during examination in transmitted light or under the microscope.

ChG are, as a rule, opaque for radiations with wavelengths less than 600 nm, i.e., they transmit only the red part of the visible spectrum range. On addition to the sulfide-selenide-arsenic systems of such heavy elements as, for example, thallium, lead-tellurium, the short-wave transmission edge shifts beyond 800 nm and glasses become absolutely opaque for wavelengths of visible range. Therefore, it appears to be impossible to define their monophase character by usual methods. The presence in the glass volume of different defects, such as bubbles and dark inclusions of vague shape, deriving from unmelting or contaminations, changes their physicochemical properties. However, the principal defect, which in some cases is impossible to fight, is the partial crystallization that causes the particular change of properties.

Thus, according to the data of Kolomiiets,[27] electric conductivity of glasses in the system As_2Se_3–As_2Te_3 at their transition to the crystalline state increased by 10^{10} times. The conductance increase at partial crystallization of glasses — initially dielectrics — belonging to the Ge–Se system is mentioned in the works by Khar'yuzov and Yevstrop'yev[28] and Borisova et al.[29]

Although the majority of ChG are highly inclined to crystallizing, the competently elaborated composition and time-temperature regime allow the producing of the monophase samples. Nevertheless, in some cases the investigators, having no idea of that, deal, in fact, not with glasses, but with glass-crystalline materials, the properties of which differ from those of stable monophase glasses.

In a large body of research of ChG properties, especially in those dating back to the first years of their investigations, the criteria used were determining the monophase character of samples, given their outward appearance, the state of the fracture surface, and, at best, the evidence from the X-ray phase analysis.

The experimental material that we have demonstrates that neither any of these attributes nor their totality allow judgment about the absence of the crystalline phase in the glass volume, to say nothing of other defects. The extreme importance of this problem makes it necessary to examine it in detail. As early as 1958 in the research of the spectral properties of glasses in the As–Se system, especially of those containing tellurium and thallium, we had established that the samples of the same composition may have different transparency, depending on the melting and time-temperature regimes. As ChG have comparatively low absorption in the region of maximum transparency, their transmission is determined by the refraction index and, as a rule, accounts for about 65 to 70% of the incident radiation (ignoring Fresnel reflection). However, for the series of samples a decrease of transmission up to the absolute opacity was noticed. No changes in the outward appearance of samples as well as in fracture surface state were noticed.

It is obvious that the evaluation of the monophase character of ChG as well as of other low-transparent materials with the naked eye is impossible.

1.2.1   X-Ray Phase Analysis as a Method for Determination of the Monophase Character of Chalcogenide Glasses

As was mentioned above, in much research on the properties of ChG, the presence of the crystalline phase is defined with the help of X-ray phase analysis. The authors of these works don't question that because of the absence of distinct lines or narrow rings on the Debye crystallogram the substance is amorphous without any crystallinity degree. Meanwhile, as long ago as the early 1960s, even those compositions in which, as we knew, there was to be partial crystallization at the taken synthesis regimes were evaluated in the literature as the amorphous ones.[26,27] To clear up this divergence we chose seven samples of glasses crystallized to different degrees and sent them to three organizations, including the Leningrad State University, to perform the X-ray phase analysis. The crystalline phase was identified only in the fully crystallized samples. It wasn't found in any of the partially crystallized glasses. After that the samples shown in Figure 2 (d to g) and a sample without crystals were sent to the laboratory headed by Poray-Koshitz in the Institute of Silicate Chemistry of the Academy of Sciences, to perform the control analysis. The analysis was carried out in vacuum chambers 57.3 mm across using monochromated CuK_{α} rays. Glass powder was stuffed into the small, thin-walled, celluloid cylinder 0.4 mm across. The regime of the pipe was power — 30 W; current — 10 mA; exposure — 100 h. The acquired radiograms are exhibited in Figure 4. There is no doubt about the accuracy of the investigation. Nevertheless, only in the glass exhibited in Figures 2 and 3(g), which was almost completely crystallized, were isolated crystals found.

To test this method's sensitivity we divided one of the samples of the monophase glass into two parts, crystallized one of them in full, then powdered both and prepared the mixture in proportions from 5/95 to 50/50. The X-ray phase analysis showed the correct proportion of crystal and amorphous parts in all samples. Thus, if the crystal phase exists separately from the vitreous one, the X-ray phase analysis precisely defines a quite small amount of it, whereas in the glass it doesn't trap even the crystallization that changes the optical properties of glasses considerably. We didn't study other physicochemical properties of the di-phase samples; however, since at the crystalline phase allocation the composition of the remaining vitreous matrix changes, it is obvious that electric, thermal, and mechanic properties must also change.

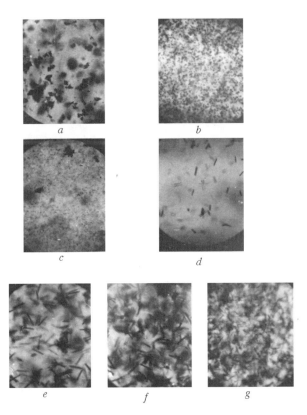

FIGURE 2. Microphotos of glass samples of different compositions with the defects of various types.

The results obtained may be explained by the peculiarities of the X-ray phase analysis, as pointed out by Brandenberger and Epprecht:[31] "At the transitions 'amorphous state-crystalline state' the presence of the interference crystalline pattern is indicative of the crystalline phase presence. The absence of such interference characteristics for crystalline bodies isn't indicative of amorphous state." Nevertheless, it is obvious that the investigators who use X-ray analysis for the evaluation of the

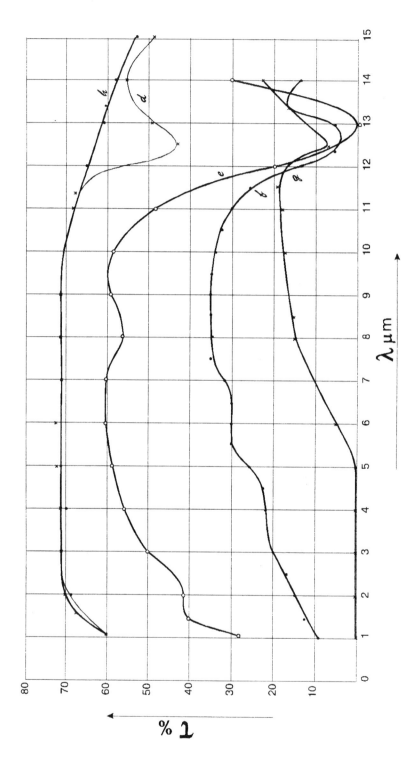

FIGURE 3. Transmission spectra of the glasses, presented in Figure 2 (d to g), in comparison with the spectrum of a monophase glass (h).

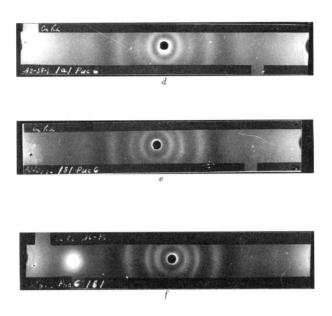

FIGURE 4. X-ray photographs of the glasses, presented in Figure 2 (d, e, f).

monophase character of ChG and who synthesize and explain their experimental material by the structural changes etc. are at risk of making a serious mistake.

We evaluated ChG quality by transmitting visible radiations with the help of a simple biological microscope, the focal distance of which increases with the removal of the upper converging lens. It allows one to examine in volume samples up to 20 mm thick at up to 400× magnification and also to photograph them with the help of a micro-camera attachment. The results of investigations of a series of transparent ChG are presented in Figure 2. It is clear that in all samples there are either dark inclusions of vague shape (a to c) or uniformly distributed crystals of various sizes (d to g). The crystallo-optical analysis demonstrated that as a rule these crystals have a short-prismatic shape and are optical-anisotropic. Their refractive index is close to that of glass. Knowing visibility dimensions and thickness of a sample, it is possible to calculate the number of crystalline and other inclusions in the glass volume by scanning from one surface to another. The number of inclusions in 1 cm^3 is determined according to the average data of six to ten planes. At on-stream scanning through the depth it is necessary to take into account only the inclusions which are precisely in the focusing plane. Because, as a rule, inclusions are not distributed evenly, the accuracy of evaluation isn't high enough, but quite sufficient for the quality evaluation and comparison of different samples. The less inclusions that are in the glass, the higher the accuracy. The sizes of inclusions are established with the help of a microscope with an integral-object micrometer (scale multiplier 10 μm). The opaqueness for visible radiation glasses is examined under the IR microscope. There, unmelted bubbles and stones against the green background of the pure glass look like dark spots of vague shape. Crystalline inclusions clearly have quite discernible cuttings and may be transparent, presenting an interference pattern in polarized light or opaque black. Spectral curves of such glasses (Figure 2) are shown in Figure 3. For comparison the spectrum of the monophase glass of similar composition is given. It is seen that the presence of the crystalline phase in negligible amounts doesn't decrease transparency and has little influence on the position of the short-wave transmission edge. When the quality and size of the crystals grow, the transmission reduces especially in the short-wave spectrum range. The spectra of the glasses, opaque under the IR microscope, look the same. The character of the change of the disperse system light transmission, determined by amount, size, and properties of the dispersion phase, is known. Thus, for example, Lecomte demonstrated that when dimensions of dust particles smeared on the transparent sublayer increases, the transmission maximum decreases and becomes more flattened.[30] The same is true for the spectral curves of the colored oxide glass with colloid coloring of golden, copper, or selenium-cadmium ruby type. Thus, the dispersion phase which appears in all cases leads to moving the short-wave transmission edge to greater wavelengths, and to a decrease in steepness.

In that way the monophase character of ChG may be evaluated according to the transmission spectra character. The criteria of the monophase character are as follows:

1. Maximum transmission is not lower than 60 to 65% according to the value of the refractive index.
2. The position of the short-wave transmission edge in the wavelength region does not go over 1.2 to 1.3 μm.
3. The steepness of the short-wave edge is from 0.1 to 0.9 τ_{max} in the wavelength interval, not overcoming 0.2 to 0.3 μm.

If at least one of these requirements is not satisfied, it points out the presence of the second phase.

Naturally, the quality of glasses, opaque under the IR microscope, can be evaluated only according to the shape of spectral curves.

1.3 GLASS-FORMING AS A FUNCTION OF THE POSITION OF ELEMENTS IN THE PERIODIC SYSTEM[3-5]

We begin with two quotations:

...On account of our familiarity with glass-ware we are apt to overlook the fact that the vitreous state is a rare physical state. Of the elements, only one, selenium, forms a glass, though plastic sulphur has some of the properties of a vitreous solid. Among compounds, the only ones which form glasses are a few oxides and oxy salts, and BeF_2.[32]

The electron-diffraction analysis shows that a series of metals has been overcooled down to the hard glass state by the precipitation of their vapours on the surface cooled by the liquefied air. Such metal glasses are quite resistant at low temperatures.[33]

At first glance it would seem that these statements are discrepant, but they reflect in the best way the fact that the glass-forming tendency depends, on one hand, on the composition that is determined by the structure-chemical peculiarities of the substances which form a glass, and, on the other hand, on the glass-transition process. In fact, applying certain technological devices and operating with small amounts of substance, it is even possible to drive metals to the vitreous state. However, these substances are mainly known in the crystalline state. At the same time the compounds which Wells reports about and also a series of composite mixtures form a glass at a comparatively slow cooling, and for some of them (e.g., B_2O_3 and As_2S_3) vitreous state is the mostly typical one and the transition to the crystalline state is difficult to achieve. In all cases when any parameter is a function of two variables, the method of their division is used when one of them becomes a constant. Accordingly, for both determination of glass-forming regions in ChG and clarification of the structural peculiarities of glasses we set the limits for the changes of glass-transition kinetic parameters and examined only the substances which transfer to the vitreous state at cooling speed not greater than 10 to 12°C/min and which withstand a repeated treatment without crystallization. Here we can only agree with Mazurin and Porai-Koshits who attract our attention to double meaning of the term "glass". It defines both a technical material and a substance in the vitreous state. The authors suggest separating the definitions "glass" and "vitreous substance".[34] To use this terminology, we'll deal in the present book precisely with glasses or, as they say nowadays, "stable glasses".

1.3.1 CLASSICAL ASPECTS OF GLASS-FORMING

If to proceed from the encyclopedic definition of structure as the totality of an object's bonds which ensure its intact and identity with its own, then for any chemical substance it should be described by the totality of the chemical bond type and spatial distribution. However, glass-forming hadn't been examined in this respect until the 1930s. In the 1920s when numerous investigations began of various physicochemical properties of glasses — thermal, mechanical, electric, optical — their abrupt changes close to the glass-transition temperature were discovered. In attempts to explain these facts numerous and quite contradictory hypotheses of glass structure had appeared. For the most significant ones, refer to the works by Tammann, the author of the popular expression, "Glass is a supercooled liquid." In 1925 Tammann had formulated the conditions of the change "liquid-glass": "Liquid passes into glass most favourably when the maximum numbers of crystallisation centres in unit of mass per unit of time are small and when the region of the greatest numbers of crystallisation centres lays in the temperature limits where crystallisation speed is very low."[35] This is undoubtedly a very interesting and practically important conclusion; that is why the so-called "Tammann's curves" still appear in the manuals for the specialists in glass investigation and manufacturing. However, Tammann's investigations didn't clear up the situation with the glass structure, because up to the present there is no universally recognized theory of the structure of liquids. The second classical hypothesis is Lebedev's conception.[36] In 1921, investigating temperature changes of the refractive index of elementary silicate glasses, Lebedev had discovered anomalous changes in the temperature range complying with quartz modification transition and he supposed that glass is the assembly of high-dispersed little crystals which in crystallography are called "crystallites". However, already in the 1960s the author of the crystallite theory didn't affirm that glass is synthesized from crystallites. He only said that in the glasses there are microregions present "with approximately regular crystalline structure."[37] However, in 1930 the crystallite hypothesis was actively supported by Randall et al. They based this upon the analysis of the curves of the intensity of X-rays scattering distribution,[38] and later the identical results of roentgenostructural analysis were reported by Valenkov and Porai-Koshits.[39] X-ray analysis demonstrated that in the glasses there are little crystals 10^{-6} to 10^{-7} cm in size. In this way the crystallite hypothesis extended very widely.

However, it must be admitted that it didn't answer the main question: why, under similar supercooling conditions, some substances change to the vitreous state and others don't. Zachariasen was the first who, in 1932, had associated the glass-forming tendency of glasses with their structure.[40]

Zachariasen didn't investigate real glasses. (We remind the reader that crystallography isn't based on real crystal research.) As applied to glass-forming he proceeded from the fact not taken into account by a lot of investigators — namely, the heating of a crystalline substance up to the temperature a bit higher than that of melting (sufficient for the transition "crystal-glass") applies to the system an impulse which is much shorter than the energy of chemical bonds. Hence their breakage is impossible in melt, i.e., global changes in the structure are also impossible and the difference in glass-forming tendency must derive from the crystalline structure of the initial substances.

At having performed a comparative analysis of the structure of oxides in crystalline state and of their glass-forming tendency, Zachariasen established that elementary compounds may be transformed to the vitreous state only in the cases when their crystalline structure represents the extended three-dimensional networks. Such structure remains in glass, but here it isn't as periodical and asymmetrical as in crystals, though not absolutely nonordered. Zachariasen considered that in nonordered vitreous networks there must exist polyhedrons from the oxygen atoms enveloping the atom A, similar to the initial crystalline networks. He formulated these conditions in four rules:

1. An oxygen atom must be connected with not more than two atom As.
2. The number of oxygen atoms enveloping an atom A must be small (three or four).
3. Oxygen polyhedrons integrate with each other in vertices, not in ribs or sides.
4. To form a three-dimensional network, at least three vertices of each polyhedron must be integrated with neighbors.

Furthermore Zachariasen divided oxides involved in the composition of glasses into glass formers and glass modifiers. To the latter he referred, first of all, oxides of the elements of the I and II groups of the periodic system which break the continuous network of bonds.

Although there are no publications on the structure of vitreous substances where Zachariasen's conditions wouldn't be listed, the author considers it necessary to cite them here, for the present-day views on the glass-forming and glass-structure problem are based on these postulates. Zachariasen's theory immediately found its supporters. In particular, Warren described his investigation of glasses by the roentgenostructural method applying the knowledge of nonordered structural network.[41] However, there were also Zachariasen's opponents. For example, Hägg blamed Zachariasen for examining glass structure by proceeding from crystals, since it was known that glass is the supercooled liquid.[42] Zachariasen objected to that, affirming that there is "no real progress to present a theory for the structure of glass in terms of the equally unknown structure of the liquids."[43] Those who criticized Zachariasen's views — mostly the supporters of the crystallite hypothesis — blamed him also for while having divided the oxides into glass formers and glass modifiers, he said nothing about the modifier's coupling with glass-forming oxides. However, Zachariasen, as well as Lebedev, didn't mention chemical bonds at all. His criteria were more likely crystallographic. Myuller was the first who had related glass-forming to a chemical bond in substances. He wrote in 1940:

> Dealing with the problem of vitreous state it's significant to know the main structural unit and the nature of its chemical bond. The glass-forming tendency of the certain substances is related to the predominance in them of the directed bonds with the reduced radius of action. These are, first of all, powerful covalent bonds between atoms of refractory oxides of the elements of the 3rd-5th groups of the periodical system.[44]

According to Myuller, the covalent component of a chemical bond is of great importance for the following reasons: electrostatic Coulomb fields which connect particles in ion crystals as well as the strength of intermolecular coupling in molecular solids are by their nature spherical and of a large action range, i.e., extending to n-particles of the given substance. Thus, in the cooling process the particles manage to compound the fully ordered structure from the melt of pure ions and molecular substances, complying with the crystalline network of a given substance at any cooling speed. A covalent bond appears as a result of generalization of a pair of electrons of adjacent atom's valene shells, its spatial arrangement in conformity with the position of the given electron orbit, and its activity doesn't concern distant particles. It prevents the system from the transition into the absolute order state at cooling and in any case it predetermines the absence of the far order.

Later, in the 1940s to 1950s, the role of a covalent bond in glass-forming had been examined by many investigators. There appeared two more glass-forming criteria, by Sun and by Rawson. Sun considered that glasses are better formed the stronger the bonds are between atoms.[45] Rawson, inferring his criterion, postulated that the execution of a continuous covalent network in melt is more possible the lower the thermal energy is at the given energy of bonds.[46] At the same time, however, works appeared where glasses were treated as the sets of ions and their structural changes were explained by applying Goldshmidt's ion radii. In this connection the author would like to cite the words of one of the leading Russian chemists — an academician, Syrkin — referring to 1946:

The fundamental factor in the glass-forming is the presence of strong directed covalent bonds which exist in fluid high-polymer. This point of view corresponds to real facts more than the attempts to manufacture glasses of multicharged ions like B^{+++}, Si^{++++} and to associate glass-forming with the ion radii correlation.[47]

As seen from the majority of publications, special attention is paid to the high strength of the covalent bonds, which in fact is typical of the bond Me–O. However, they are much weaker in, for example, chalcogenides (Si–O; 105 kcal/mol, As–S; 49 kcal/mol), for their glass-forming capacity isn't lower. Hence the energy of bond breakage isn't of great importance in the glass-forming.

1.3.2 THE ROLE OF CHALCOGENIDE GLASSES IN CLARIFICATION OF THE VITREOUS STATE NATURE

First of all, the investigation of chalcogenide glasses established that the principal mechanisms which relate the glass-forming tendency of substances to composition and structure are traced in them much easier. Really, even the simplest oxide systems, such as SiO_2–R_2O_3, SiO_2–R_2O, and RO (where R = metals of the I, II, III, and V groups), contain atoms of three types. Thus, the introduction of the elements of the first and second groups leads to the appearance of ion bonds which causes the formation of polar structural elements in nonpolar medium, leading to association, hence to the emergence of structural microheterogeneity (after Myuller).[48] Unlike oxide systems, chalcogenide ones can consist of only two atoms (As–S, As–Se, Ge–Se) at considerable composition change. In addition, in elementary, binary, and ternary systems as well as in complex ones, atoms, as a rule, are connected only by covalent bonds. That enables one to evaluate how the glass-forming tendency changes with a composition, proceeding from structures of valent shells of atomic components, with no regard to polarization and electrostatic forces.

Later, as a result of oxide glasses investigations, the hypotheses appeared which postulated the necessity of mixed couplings for glass-forming. Thus, Smekal considered that covalent-correlated substances can't form glasses without bond ionization to a high degree (Si, B unlike SiO_2, B_2O_3). He also pointed to the fact that in selenium with a chain structure there exist mixed couplings covalent in chains and van der Waals between them.[49] Stanworth also carried out the correlation between the glass-forming tendency and bond ionization degree.[50] Meanwhile, in the system As–Ge–Se where the greater glass-forming region is observed, bonds have a very low degree of ionization. To calculate after Pauling, then the ionization in the bond As–Se, due to the adjacent position of the atoms in the periodic system, is about 6% and the ionization of the bond Ge–Se is about 12%. Besides, in a certain part of this system, for example, along the cut As_2Se_3–$GeSe_2$, ternary network of bonds is performed (as in SiO_2) where there are no forces of chain-to-chain or layer-to-layer interaction. It is quite obvious that the presence of mixed couplings isn't necessary for the glass formation. The investigation of the glass-forming tendency among oxides and chalcogenides enabled Winter-Klein to relate this property of substances to the outer electron configuration of atoms, hence to the element's position in the periodic system.[51] She paid attention to the fact that all the elements which form a glass on their own or take part in vitreous network formation have p-electrons in the outer shell; of primary importance here are the atoms which contain four outer p-electrons, i.e., atoms of the elements of the sixth group of the periodic system.

Though the author doesn't explain the cause of the four p-elements' special purpose, the very statement of the problem of the glass-forming association with the periodic law was very progressive.

The investigation of the glass-forming tendency among compounds and mixtures based on chalcogens containing 27 chemical elements allowed us to establish that the greatest number of glasses and binary and ternary systems is formed by combination of the elements of the fourth to seventh principal subgroups (Table 1). It is obvious that all these elements obey the well-known rule in chemistry, derived by Hume-Rothery in 1930, also known as "the rule 8-N", which says: "In the crystals of the elements of the 4th–7th groups each atom has 8-N nearest neighbors, where N is the ordinal number of the periodic group."[52] It should be emphasized that 8-N indicates the number of unpaired electrons — therefore the number of possible valent bonds. Coulson writes: "It will be recognised that this is merely another version of the octet rule of molecular chemistry. The applicability of this rule to a crystalline solid may be regarded as strong evidence that the bonds are localized ones, of molecular-valence character."[53] Mott points to "the rule 8-N" applicability to ChG.[54]

TABLE 1
The Elements Used in Our Work

	I	II	III	IV	V	VI	VII	VIII
1								
2			B*					
3			Al*	Si*	P*	S*		
4	Ca⁻	Zn⁻	Ga⁺	Ge⁺	Ti⁻ As⁺	Se⁺	Cr⁻	Mn⁻ Fe⁻
5	Cd⁻		In⁺	Sn⁺	Sb⁺	Te⁺	Mo⁻	
6	Ba⁻	Hg⁻	Tl⁺	Pb⁺	Bi⁺	W⁻		

⁺These elements form glasses
⁻Non-glass-forming elements
*Can form glasses, but of low quality

The fact that elementary glasses are formed from the elements which obey the rule by Hume-Rothery is definitely indicative of the correspondence between a coordinating number and the atom's valence necessity for glass formation, i.e., of the covalence structure correlation. To accept the above-cited observations without any limits, we should expect the similar glass-forming tendency in all the substances, correlated by covalent bonds, but the fact is that among chalcogenides of the elements of the fourth to fifth groups the glass-forming tendency varies. Therefore, covalent correlation is a necessary but not sufficient condition of glass-forming.

To determine other criteria it should be expedient to examine elementary compounds and to explain the causes of presence or absence of glasses among them, with regard to the position of component elements in the periodic system and comparing crystalline structures of corresponding substances. From this viewpoint the difference in the glass-forming tendency of the elements of the sixth group attracts attention first. According to the valene shell structure ($S^2P_x^2P_yP_z$) all the elements of the sixth group form, in normal conditions, two bonds at the expense of "idle" p-electrons and their coordinating number is, as a rule, equal to 2. If an adjacent atom has a free orbit, it can form a donor-acceptor bond with the help of an unshared pair of electrons which, as we'll see further on, greatly influences the glass-forming ability of the substance. It is known, however, that in spite of electron configuration similarity the structures of the elements of the sixth group as elementary substances are different. Oxygen as well as nitrogen doesn't obey the rule by Hume-Rothery: the number of its nearest neighbors is equal to 1. On solidification oxygen forms molecular crystals and on melting down — liquid, consisting of the molecules O_2 and O_3. All attempts to get vitreous oxygen were unsuccessful.

Structures of sulfur, selenium, and tellurium — both in solid and molten states — had been studied in the 1970 and 1980s by a series of physical methods due to their semiconductor properties. The detailed review is presented in the monograph by Feltz (3.1.3).[8] The structural chemistry of these elements is defined by the formation of covalent bonds between each atom and two adjacent ones. However, selenium and sulfur differ greatly in thermodynamic stability of their structural forms: the stable form for sulfur is a molecular structure of rings with a different number of atoms (from 3 to 20); and for selenium the stable structure is that of the trigonal modification representing spiral chains. Three other selenium modifications, viz., α, β, γ-monoclinic selenium, consisting of Se_8 rings, are metastable and at fusion they change to the trigonal modification.

During melting the structures of both selenium and sulfur change continuously with temperature change. At a certain temperature (432.5 K) viscosity of the sulfur melt increases irregularly in more than 5 orders that can be explained by the rings' breakage and formation of chains with an average length of 10^5 atoms. The maximum viscosity is attained at 460 K and after that it starts to decrease.

Vitreous sulfur is produced at the abrupt cooling of melt exactly in this temperature interval. Sulfur also forms glasses in binary systems, although in a comparatively narrow composition range with a slight deviation from stoichiometry. All the attempts to manufacture stable glasses with a sulfur excess were of no result.

As already mentioned, selenium as a glass-former occupies a peculiar place among the elements of the periodic system. It exists as a stable glass and can easily form a series of binary vitreous systems with its subgroup analogues as well as with the elements of the groups IV and V, and also forms complex glasses in a wide compositional range. In the literature there is no consensus on both the structure of vitreous selenium and the structure of its melt. In this structure can be found octanomial rings, combinations of rings and chains, and just chains of various lengths.

Let's examine the structure of selenium melt, taking into account the experimental data which have come to hand by physical methods. The structure of selenium melt should apparently depend not only on temperature, but also on the initial crystal structure. Suppose that at the melting of monoclinic selenium modifications, transfer to the trigonal modification with octanomial rings' breakage doesn't occur instantaneously, therefore at a certain temperature interval the melt represents the mixture of rings and chains and this very structure gets "frozen" in glass. As the temperature rises further, the melt fully polymerizes into the spiral chains with the distance between them more than that in crystals. At the fusion of trigonal selenium with initial chain structure, it (the structure) is preserved close to the evaporation temperature. Only the length of chains may change.

We suppose that the results obtained by investigators of the vitreous selenium structures were so different due to the selenium polymorphic modifications, because they depend on the temperature conditions necessary for the preparation of samples. It's important for us that the very difference between sulfur and selenium in the glass-forming tendency leads to the conclusion about the polymer structure of glasses.

Many examples exist evident of the fact that molecular substances don't transfer to the vitreous state. Consider the most characteristic ones. Aluminum oxide (aluminum-boron's analogue in the subgroup; occupies the position in the same period with arsenic and silicon) is crystallized in the form of covalent-correlated closed molecules Al_4O_6 which are preserved in melt. Aluminum oxide forms a part of silicon and other complex glasses, but it doesn't form glasses by itself at any accessible cooling speed.

At first sight the situation with arsenic oxide is absolutely different. As_2O_3, known as arsenolite, is formed, as a rule, by precipitation of a solution and is crystallized like aluminum oxide in the form consisting of molecules As_4O_6. Nevertheless, at heating over 430°C and insufficient supercooling, As_2O_3 forms the stable glass. In particular, this fact provided grounds for a group of investigators to make a conclusion about the molecular structure of glasses.[55] However, at T = 426°C arsenalite transfers to its polymer modification — claudetit — which is characterized by a laminated network, and As_2O_3 in the vitreous state can be manufactured only in the temperature conditions which provide polymer modification presence in melt. We should note that As_4O_6 molecules were not discovered by physical methods at the investigation of the vitreous As_2O_3 structure. KR spectra and KRR are interpreted on the basis of the claudetit structure. Polymer modifications for phosphorus oxide are also known.

On the structural diagram of the system As–S two congruently melting compounds are marked: As_2S_3, orpiment and As_2S_2, realgar. Orpiment occurs in the crystal state, but when supercooled down to the glass it is crystallized hard. Trigonal structural groups in it are fixed in a two-dimensional lattice which, like claudetit, consists of 12 nominal corrugated rings. Realgar, unlike orpiment, isn't produced in glass form. It consists of As_4S_4 molecules where arsenic atoms positions are in tetrahedron's vertices and sulfur atoms' positions are close to the flat square. Glasses are dealt with as polymers in the works by Kobeko,[33] Hägg,[42] Weyl,[56] Stewels,[57] Tarasov,[58] and others. Zachariasen and Myuller examined glasses only as polymers. In the beginning of the 1970s it seemed that the polymerism of glasses was proven. Nevertheless, along with the numerous investigations of glass structure corroborating this point, models of Phillips molecular clusters,[59] Hoseman's microcell crystals,[60] and others have appeared in the last decades. In some works the amorphous structure is simulated from clusters by computers or is described as the system of particles, i.e., little crystals, for example, in the Pinsker model.[61] Rollo and Burns, in 1991, wrote about molecular inorganic glasses,[62] based on evidence from the structure of P_4Se_3, which isn't obtained as stable glass. In our opinion the appearance of the molecular model is, up to now, based on the existing confusion of the knowledge about amorphous and vitreous states. Some investigators identify amorphous and vitreous states; others state a difference between them, not understanding clearly how these forms of matter vary in structure.[63] Meanwhile, the majority of amorphous substances represent (after Randall[38]) "crystal-like formations where crystalline grains are extremely small and unevenly spreaded;" the distinction of the vitreous state as a special form of the amorphous state is, in our opinion, exactly by the fact that glasses represent chain, laminated, or three-dimensional polymers. However, the condition of polymerism, as well as of covalent correlation, is necessary but insufficient for glass-forming, i.e., if all glasses are polymers, not all polymers are glasses.

It is known that the capacity for glass-forming decreases with the increasing of the elements' atomic numbers. That is why the glass-forming criteria, including that by Winter-Klein, don't "work" if this fact is ignored. Kolomiiets and Goryunova, who were the first to discover this peculiarity in the series of arsenic chalcogenides, explained it as the covalent bond "metallization" and we consider it to be rather vague.

Let us examine this phenomenon, proceeding from the information obtained by physical methods, about the structure of crystalline forms and melts of the chalcogenides of heavy elements, — as we had done earlier with sulfur and selenium. The glass-forming tendency among the elements of the sixth group enhances sequentially in the series $O \rightarrow S \rightarrow Se$ and decreases abruptly in tellurium. Tellurium was never met in the form of stable glass. It forms neither binary systems like selenium nor even vitreous compounds with stoichiometrical composition like oxygen and sulfur, in similar cooling conditions, although it is crystallized in the same structural type as trigonal selenium. However, the structure of melted tellurium, examined with the help of neutron and X-ray diffraction, differs from the crystal one in coordination: instead of two shorter and four longer distances between atoms, there are registered in melt three closest and three farther atoms, i.e., this structure is close to the one of rhombohedral arsenic. Ge Te is also crystallized in the structural type of rhombohedral arsenic (distorted NaCl structure).

Crystalline As_2Te_3 differs in structure from As_2Se_3 and As_2S_3. In telluride the coordinate numbers of arsenic atoms are six and three and they are placed so that they form bands from polyhedrons $AsTe_6$ and $AsTe_3$. One can see that in tellurium, like in its compounds, the coordinate numbers are higher than in arsenic and germanium sulfides and selenides. Arsenic telluride and germanium telluride don't transfer to vitreous state under the cooling conditions that we have accepted.

Germanium analogues in the subgroup tin and lead, as well as arsenic analogues, antimony and bismuth, as compounds of stoichiometrical composition don't form stable glass with any chalcogen. It is known that chalcogenides of tin and lead are crystallized mostly in a sodium chloride network (coordination 6:6). In general large coordinating numbers are typical of all heavy elements' compounds. This phenomenon, after Krebs,[64] is explained by the fact that due to the tendency to hold an unshared pair of S-electrons, atoms of heavy metals lose the ability to sp^3 hybridization and form bonds mostly at the expense of pure p-orbits. As the edge surfaces of p-orbits are symmetrical about the nucleus, the atoms are enabled to form bonds in a reverse direction, thus increasing the number of the closest neighbors. Anyway, our analysis corroborates in general the second postulate after Zachariasen: "the number of oxygen atoms around an atom A shouldn't be large. (3,4)" However, such elements as Si, Ge, or As, with crystalline structures representing covalent-correlated polymers with coordinating numbers not larger than four, easily change to the amorphous state but are not met as glasses.

Hence one more condition for substances' transition to the vitreous state is necessary. Proceeding from the above-described sulfur and selenium structures it is obvious that in the vitreous state they represent curved chains, i.e., each atom has only two closest neighbors. Such bonds are called "bridge" bonds and Myuller continued to discuss their meaning in glass-forming. In particular, he formulated the following explanation of the peculiar role of the elements of group VI in glass-forming: "This group is the only one of the kind where atoms of elements correlate through covalent bonds with only two adjacent atoms — thus being involved in the glass network in the form of units forming interlacements."[65] As a rule these elements form only two covalent bonds in the compounds with the elements of groups 3 to 5 as well.

The knowledge of the glass-polymer structure and of these bonds' role in glass forming is presented most completely in the works by Tarasov, dating back to 1959 to 1965.[58,66,67] He wrote, in particular:

> ... the polymers of this type, along with ion fractions of bonds, have in their shells covalent fractures of bonds of sp^3 type (i.e. O–Si–O bonds) — inside tetrahedrons and the bonds close to the pure p-bonds (Si–O–Si bonds) — between tetrahedrons. Such duality is important for understanding of the glass origin on silicon and germanium oxide. The fact is that tetrahedral covalent sp^3 bonds are rather tense, i.e. their deformation energy is high not only on valent distances but also in tetrahedral valent angles.

And further on:

> The bonds like Si–O–Si or Ge–O–Ge have another structure of wave functions. They are like hinges, hence the structure of such glass-formers is easily deformed due to variations of Si–O–Si angles. These considerations help to understand the low glass-forming tendency in the structures where the frames are constructed on the bonds of sp^3 type only (structures of diamond, wurtzite, sphalerite).[58]

It can be seen, however, that the presence of bridge bonds doesn't lead to glass-forming. In particular, the "bridges" exist in present in palladium halogenides with the general formula

(Str. 1)

and in diboranes. Neither of them is known in the vitreous state. Silicon disulfide is formed in much the same way: every two silicon atoms are combined through the sulfur atom into the polymer band structure of the type

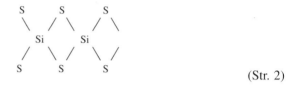

(Str. 2)

Silicon disulfide, as opposed to SiO_2, does not form glass in standard conditions. Pauling believes that SiO_2 and SiS_2 structures are different due to the amplification of the covalent character of the bond Si–S (Si–O — 50% ionization; Si–S — 30%), because the less ionized sulfur atoms, as opposed to oxygen atoms, don't tend to repel on the maximum distances and can fix every two silicon atoms by two bridges.[68] As this example shows, the amplification of a covalent component of a chemical bond can influence the glass-forming, even unfavorably as well. This structure can also be called "hinged", as it allows the angles to curve in two directions. However, it is insufficient for the glass-forming: the bond of the "Hook's hinge" type is necessary here. Such rarely met structure is met here, which provides not only for the presence of two closest neighbors of the coupling element (in the given case, chalcogen), but also for combining of the adjacent atoms through only one coupling element. Only such a connection of structural polyhedrons enables bond angles to change in all directions and thus ensures the disturbance of the far order.

This statement, in a concealed way, is presented in Zachariasen's postulate which requires that all polyhedrons be jointed only in vertices.

All of the above-examined considerations, based on the investigations of glass-forming and physicochemical properties of glasses, are verified by later structural researches, performed by different authors with the help of various physical methods. Thus, Feltz cites evidence that the valent angles between the bonds Si–O–Si change from 120° to 180°. There are variations of angles between the bonds As–X–As and Ge–X–Ge, where X is sulfur and selenium, as well as in selenium itself, and some changes of their lengths are also established in the structural investigations (2.4.2.2).[8]

There's no necessity to introduce the idea of each peculiar bond, i.e., the two-center bond after Dembovsky and Chechetkina[20] or π_{d-p} and δ_{d-p} chalcogenide bonds put forward by Funtikov.[69] A hinged bond, remaining space-directional and short-range, affords the flexibility necessary for the distortion of a crystalline network.

The third condition for the glass-forming is formulated as follows: glasses can be formed only when, in the melts of initial substances, there are bonds present which allow angles to change in all directions.

Thus, having examined glass-forming as the function of the elements' position in the periodic system, we can come to the conclusion that the following necessary and sufficient conditions exist for glass-forming or the principal peculiarities of the vitreous state:

1. The localized twin-electron bonds are present in the structure.
2. The structural network consists of infinite polymer complexes.
3. The adjacent complexes are connected through only one bridge bond, i.e., the bonds are present in structure which can be called hinged bonds.

That which we call "complexes", Zachariasen "polyhedrons", and Myuller "structural nodes" or "structural units" are nothing more than central atoms with their first coordinating sphere or the aggregation of atoms, the multiple occurrence of which allows the given vitreous network to form. This also can be called "the regions of near order".

In our opinion, everything said above is the only generalization which applies to the vitreous state as a whole. The structure of glasses of various compositions can be absolutely different within the indicated limits. The diverse polymerization degree may take place here. In parallel with the covalent bond, obligatory for the formation of the vitreous network, there can exist other bond types, such as ion, van der Waals, or coordinating-valent ones. In glasses containing ionic components, microheterogeneity after Myuller is possible. We had come to these conclusions in 1969. Nowadays they could have been stated in better terminology, however, they are the same in essence at present. It's easy to notice that, hence, some certain consequences follow, to wit: elements or elementary compounds may transfer to the vitreous state only if all the signs of a glass-forming structure are preserved in their melts, because in this case the glass structure can't differ considerably from the melt structure.

We should notice that among all the elements of the periodic system only selenium answers these requirements:

- In complex systems which consist of several elements or elementary compounds, glass-forming structure may derive from mutual distortion of two or more nonglass-forming structures. That especially applies to the elements of big periods, as the absence of hinged bonds is typical of the crystalline forms of their compounds. The combination of heterogeneous structures leads to the changes in the first coordination sphere of atoms and, as a result, coordinating numbers diminish and the hinged straps are formed, i.e., there occurs an opportunity to form a glass.
- In complex glasses there may take place structural units with nonglass-forming structures, if they are surrounded with glass-forming complexes.

- Both the energy of a chemical bond and the ionization degree don't influence the glass-forming — it is sufficient for a bond to be spatially localized and directed.
- The glass-forming isn't connected with either presence or absence in structure of the comparative geometric crystallites' order or, in modern terminology, of the middle-order range. The near-order ranges discovered currently account for 5 to 7 Å, i.e., they approximately correspond to the size of one silicon-oxygen tetrahedron.

In 1987 Elliott[70] and also Price and Susman[71] advanced the evidence of the presence of middle-range-order (MRO) regions in the structure of vitreous substances, in particular, of ChG; and they were, undoubtedly, right, for the existence of such regions is inevitable in real glasses. Really, taking into account that any system tends to the minimum of free energy, one can't imagine the disordering of a melt's structure up to separate polyhedrons, to say nothing about its preservation in this state at real cooling rates. However, the glass nature isn't connected with the "middle-order" existence, for even its complete absence can't make the vitreous state disappear.

Nevertheless, we shouldn't take into consideration the possibility of relative geometrical ordering, even in real glasses, unless their properties don't depend on it.

Here, in fact, lies the disagreement in Zachariasen's and Lebedev's hypotheses. No doubt they both are right, especially if one takes into account that the author of the crystallite hypothesis himself mentioned the extraordinary small dimensions of the ordered ranges in the distorted network.[37]

In real glasses as well as in real crystals various defects are possible. Probably, they can occur even more frequently in crystals, for such defects as blocks, dislocations, and broken bonds must be "curved" in the melt. However, the defects of real crystals are not taken into account in the description of their structures. At the same time, in some publications on glass structure particular importance is attached to point defects, i.e., to the bond breakages that are regarded as the necessary glass-forming condition.

It's quite natural that the more sensitive the methods for the substance structure analysis will appear in the future, the more subtle the details of the glass structure will be discovered by investigators. In particular, the presence of molecular units not connected with the principal network, e.g., short chains or closed rings, as well as of the fragments of a nonglass-forming structure, blocked in the glass-forming one, is possible in any glass, but, like the middle order, it isn't connected with the nature of the vitreous state.

Just as crystallography operates with ideal crystals — and the fundamental gas laws derive from ideal gases — in the glass science, while establishing the peculiarities of the vitreous state and glass structure, one should proceed with the knowledge of ideal glass, not taking into account the defects, typical of real materials. As to ideal glasses, they are covalent-correlated polymers which consist of static, evenly spread polyhedrons with fully saturated valent bonds, an obligatory presence of bonds which easily change their position in space, and that predetermines the absence of the far order.

1.4 GLASS-FORMING, PHYSICOCHEMICAL PROPERTIES OF CHALCOGENIDE GLASSES AND THEIR STRUCTURAL MODEL[6-17]

In all our works we, as a rule, carry out the substitution of components — partial or complete — by their analogues in the subgroup which allows us to complicate compositions with the least structural distortions.

It should be emphasized that the compositions of glasses are calculated only as a percentage of atoms. Even when sulfur is used in a persistent state, as arsenic trisulfide, arsenic and sulfur at compositions' calculation are treated as an independent element in continuous ratio.

Such a method of compositions' calculation provides the only opportunity to analyze systematically the influence of various elements on glass-forming and the properties of glasses.

Complete-enough information on the property change with composition can be obtained during the synthesis of glasses in at most 5 at.% of each component of the system. In the region of abrupt changes of properties the difference mustn't be greater than 1 to 2 at.%. Due to the practical orientation of this work, we study only those properties of glasses which are necessary for their application as optical materials. At the same time they all enable us to make judgments about the substance structure. Of great interest here is the density, which is pointed out by different authors. Thus, for example, Barret writes:

> If the liquid density doesn't differ a lot from the density of a solid, we can't suppose the difference between crystalline and liquid states in co-ordination or in distances from the nearest neighbours, therefore the precise measurements of density are almost as valuable for the judgements about the substance structure as the best X-ray analysis.[72]

It's impossible, however, to use the glass density values at the analysis of the composition-structural change, for this change gets diluted through the difference in atomic weights of the introduced and removed components. To eliminate the difference in atomic weights, Yefimov suggested expressing the density as the total atomic concentration in a unit of glass volume. Recalculation is performed by the formula: $N = \rho / \overline{A}$, where N is the total concentration of atoms in 1 cm^3 of glass; ρ is glass density; $\overline{A}$ is average atomic weight of glass. Introduction of the notion "average atomic glass weight" is necessary for ChG, because they, as a rule, are synthesized from the elements in nonstoichiometric relations. $\overline{A} = \Sigma X_i A_i$, where $X_i A_i$ are atomic fractions and atomic weights of the elements entering into the glass composition. Applying the total volume concentration instead of the density it is possible to find the real change in atom packing.

Besides density, the most substantial changes in structure influence such structure-insensitive property as microhardness, and the test for it represents one of the numerous ways of evaluation of the mechanical strength of substances. The experiment has demonstrated that microhardness of ChG depends linearly on the concentration of atoms. With the change of glass composition it undergoes bends in the points where N extremes are observed.

As is known, viscosity is rather sensitive to structural changes. In particular, Yevstrop'yev demonstrated it for oxide glasses.[73] Unlike density and microhardness, viscosity represents kinetic characteristics which describe the process of the transition "melt — hard glass" and vice versa. Therefore the viscosity values at a given temperature are estimated by the energy parameters of viscous flow, i.e., by energy and entropy of activation. In the SI system the viscosity is expressed in Paxs, and in the CGS system g/emxs unit called poise (P); $1P = 0$, 1 Paxs. In the works 60–70s we expresss viscosity by the CGS units. Nemilov has stated that the energy of activation at the viscosity value 10^{15} to 10^{16} P is close to the energy of a chemical bond and is in certain interrelations with the glass structure.[74,78] In this way, the coincidence of temperature trend of viscosity for the glasses of various compositions in proximity of the chemical bonds' energy of the components points out the uniformity of these glasses' structures. On the contrary, the difference in temperature trend of viscosity is evidence of the difference in their structure.

Having an idea of viscosity change with the temperature, we can organize the correct technological process of glass manufacturing. To certain viscosity values there correspond the temperature values, characterizing the principal thermal properties of glasses: glass transition temperature, softening point, annealing temperature, upper use temperature.

These values can be obtained also by the dilatometrical method. Besides, dilatometry enables one to calculate the thermal coefficient of linear expansion. In Figure 5 one can see the typical dilatometrical curve of the glass of the As–Se system. The position of glass-transition temperature (T_g) and the highest point on the dilatometry curve — dilatometrical softening temperature (T_{ds}) — are shown. Coefficient of thermal expansion (α) is calculated from the formula:

$$\alpha = \frac{\Delta L}{L \Delta T} \cdot C^{-1}$$

Our experience shows that for the absolute majority of ChG T_g corresponds to viscosity 10^{13} P, and T_{ds} 10^{11} to 10^5 P. It should be noticed that T_{ds} is more important at application than T_g, for it enables one to determine the upper annealing edge more accurately. T_{ds}, as well as T_g, is very sensitive to structural transformations. Such practically important properties as chemical and radiation stability are quite good in ChG and change but only a bit with the composition.

In Figure 6 the combinations of elements are presented, from which the glasses investigated by us were produced. The elements, which enter the glass in the amount not exceeding 10 at.%, are underlined. The changes of properties in all the systems are connected with the structural transformations. The obtained results have provided the basis for the structural model of ChG, which accounts for the view of glasses as the covalent-correlated polymers, elaborated on in the previous section.

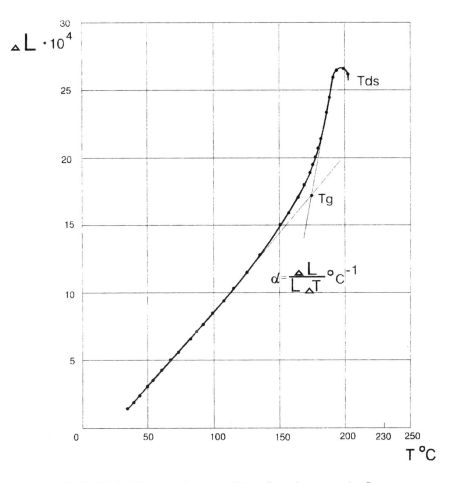

FIGURE 5. Dilatometric curve of glass from the system As–Se.

1.4.1 PREPARATION OF SAMPLES AND INVESTIGATIVE TECHNIQUES

To determine glass-forming regions and investigate the properties, glasses are synthesized according to the above-described technique. The batches are of 10 to 50 g. After the synthesis process is over, the ingots, removed from the melting ampules, are mounted in a mold and put through molling and annealing. The time-temperature regime of molling and annealing is set for each group of glasses, respectively. The annealed samples are checked for their monophase character and, proceeding from this, one can draw inferences about the boundaries of glass-forming regions and possibilities of the properties' measurements. The sample quality is checked by viewing under the microscope in visible and infrared spectrum ranges and also by comparison with the standard spectra for each glass type.

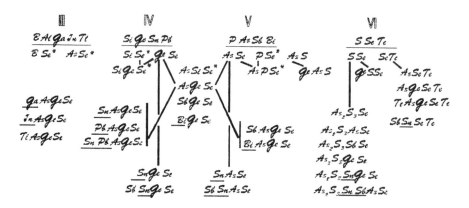

FIGURE 6. The systems investigated in the SOI.

The properties of glasses are determined by the universally accepted techniques on the samples with no three-dimensional defects or with defects, the number of which can't influence spectral transmission.

The density is measured by hydrostatic weightings in toluene (methylbenzene) with a precision of 1×10^{-3}; the measurements are carried out on at least two samples of each composition, annealed in one and the same regime. Among various methods and devices for microhardness tests we've chosen the pyramidal microhardness meter PMT-3.[75] The difference between this technique and the measurements after Vickers is that the load is constant and a diamond pyramid is used for indentation. The size of its print, estimated according to the sizes of the diagonal with the help of an ocular micrometer, serves as the measure of microhardness. In determination of the imprint's diagonal length lies the main source of errors which can appear at operations at this device. The maximum error at two measurements may be as much as 3%. To enhance the precision of the method, the size of a diagonal is determined for not less than five prints on each sample, in control cases by two operators. The per-unit error then reduces to 1 to 1.5%, which is quite sufficient with our purposes. We perform the measurements under a load of 50 g. The measurements of ChG T_{ds} can be performed by different methods, in which either a sample's bend or its change at warming up under load is used. Thus, Hilton and others propose to determine T_{ds} with the help of the device, in which a flat-parallel disk is placed on the hollow and it buckles at the warming up under the constant load.[130] Savage and Nielsen applied the method in which a glass rod of tetragonal section 1 mm thick buckles at temperature rise at a speed of 1 to 2°C/min by gravity.[131] Kolomiietz and Shilo determined the softening point by the method, based on the measurement of the depth of poisson's immersion into a sample under the effect of constant power 1 kg/min at various temperatures.[132] Pearson and others measured the initial change of the sample size under the influence of a temporary load.[114]

We measured T_{ds} and calculated α with the help of a quartz vertical dilatometer constructed by Sorkin.[76] The precision of the device for α measurement is 2 to 3 $\times$

$10^{-7}°C^{-1}$; for T_{ds} 3 to 5°C. If the control measurements are necessary, the interferometer method is applied. For fusible glasses we have worked out and now use the special method, described below (see Section 1.4.5.2). The glasses' viscosity can be measured by various methods, namely, caving in or pressing in of a bar, twisting of a rod, etc. A very popular method is a thread stretching, which allows one to perform measurements in the region 10^{15} to 10^9 P and gives the absolute magnitude of viscosity. We had been working on the viscosimeter constructed by Slavyanski, pressing in the indentor.[77] This method, modernized by Nemilov and Petrovsky, is based on the determination of the speed of the cylindrical indenter impression into the glass plate.[133] The measurements are performed in the dynamic regime at a sample heating speed of 1 to 2°C/min. As an indenter a quartz fillet with a small ball on an end is used. The viscosity values being measured range from 10^{13} to 10^4 P. Here, like in the tension method, the constant impression speed is achieved. Viscosity is calculated from the formula:

$$\eta = A\tau_{100}P$$

where τ_{100} is constant time of indenter's impression, observed during the experiment on 100 graduations of ocular micrometer scale; P is the loading; A is the constant of the device, to determine which graduation on the glass with different viscosity is necessary. To measure viscosity by this method, no large amounts of glass and fine treatment of samples' surfaces are needed. The accuracy of the method is ±0.061 gh.

It should be emphasized here that the comparatively high reproducibility noted at the investigation of ChG can be explained by the peculiarities of the synthesis technology. The results of the measurements of oxide glasses of distinct melting properties can be influenced by the distinctions in melting atmosphere, dissimilar degree of the component volatility determined by melting conditions, etc. These factors in some cases are related to greater or less discrepancies between true and calculated compositions. The synthesis of oxygen-free glasses is carried out in evacuated space with the primary factors, which determine the process, being permanent. As a result, practically no compositional changes occur.

1.4.2 GLASS-FORMING IN THE SYSTEMS BASED ON SELENIUM WITH THE ELEMENTS OF THE SECOND AND THIRD PERIODS (P, Si, B, Al)

Among the elements of the second period phosphorus gives binary selenium glasses in the composition region from pure selenium to $P_{60}Se_{40}$. However, already on introduction of 10% of phosphorus the glasses become chemically unstable due to the interaction with water vapors in the atmosphere by the reaction: $P_2Se_3 + 3H_2O \rightarrow P_2O_3 + 3H_2Se$. Also noted is a high crystallization tendency in these glasses.

At a sequential phosphorus replacement with arsenic in the binary system P–Se, one can obtain vitreous substances over the whole composition range — from $P_{60}Se_{40}$

to $As_{60}Se_{40}$. When the phosphorus content is more than 40 at.%, glass-forming seems to enhance slightly: arsenic and phosphorus are injected as a mixture at a rate of 70 to 75 at.%. However, in this composition range the glasses are, as a rule, removed from the melting ampul with the explosion and they tend to self-ignite. On the walls of an ampul one can notice a white deposit which spreads out into the air. This phenomenon is apparently connected with the partial phosphorus evaporation in the synthesis process.

Stable glasses which don't decompose in the air and have low crystallizability are produced only with phosphorus content at the rate of not more than 5 at.%. The properties of the phosphorus glasses at that don't differ a lot from the ones with arsenic, i.e., the substitution of arsenic with phosphorus is of no practical interest. In the binary system Si–Se the glass-forming region is not nearly so large as in the arsenic one; glasses are formed to the composition $Si_{20}Se_{80}$ but, like the phosphorus ones, they rupture on exposure to the air and the released H_2Se — poisonous gas with a sharp odor — gives no way to the investigation of their properties.

In the system As–Si–Se, even at the melting temperature of 1100°C, homogeneous glasses have never been obtained. All the samples involved plenty of dark, irregularly shaped insertions. The nature of these insertions as well as their number increase as the synthesis temperature reduces, and their independence from the cooling regime causes one to suppose that it is not crystallization but poor penetration which takes place here. The temperature of 1100°C is apparently insufficient for the reaction of silicon with other components. If the temperature rises in quartz ampules in the melting process, it may cause their disrupture in the furnace. Besides, the arsenic addition only slightly upgrades the chemical tolerance of the silicon-selenium glasses: in a ternary system as well as in a binary one the majority of samples rupture by the action of the atmosphere moisture.

The attempts to insert aluminum and boron into the arsenic-selenium glasses have also failed. The supplementary germanium addition can't strengthen the systems with these metals as well. In none of the cases are good glasses produced: there is no homogeneity in vitreous masses and, above all, the materials in use are rather hygroscopic and rupture quickly with the release of H_2S and H_2Se.

Similar results are presented in the works by Hilton and others.[79,80] The most tolerant glasses were obtained by the authors in the systems Si–As–Te and Si–Ge–As–Te. However, they have drastic extinction coefficients and it points out the presence of the second phase. Savage and Nielsen also studied the glasses in the Si–As–Te system and they have come to the conclusion that the materials obtained contained the crystalline phase which could deteriorate optical parameters of components manufactured from them.[81]

It is evident from these experiments that the elements of the second and third periods, namely, B, Al, P, Si — which the oxygen glasses are based on — don't form stable glasses in the sulfoselenide systems. As a consequence they were pronounced unfit as the components of the optical oxygen-free glasses and were never used in further elaborations.

1.4.3   GLASS-FORMING PROPERTIES AND STRUCTURE OF GLASSES IN THE SYSTEMS CONTAINING THE ELEMENTS OF THE IV TO VI MAIN SUBGROUPS OF THE FOURTH TO SIXTH PERIODS

1.4.3.1   Glass-Forming in the Systems As–Ge–Se, Sb–Ge–Se, Bi–Ge–Se

Combinations of the elements of the periods 4 to 6 of the groups IV to VI in the periodic system provide the greatest variability of the stable oxygen-free glasses. The largest glass-forming region was found in the system As–Ge–Se which we had first studied in 1959 to 1960 in collaboration with L. G. Aijo. At the same time in the Physico-Technical Institute several compositions were obtained along the pseudobinary cuts $As_2 Se_3$–$GeSe_2$ and $As_2 Se_3$–$GeSe$. In that work, however, the absence of glass-forming in the binary system Ge–Se was pointed out and the properties of glasses were not studied at all.[82]

In 1962 to 1964 this system was investigated by Baydakov[83] and Nielsen,[84] and a bit later by Haisty,[85] Webber and Savage,[86] and others. The glass-forming in the system As–Ge–Se is distinctive in that here the great amount of glasses is formed beyond the line of the pseudobinary cut $As_2 Se_3$–$GeSe_2$ and even in the region where the overall content of arsenic and germanium is considerably more than that of selenium. The same is true for the binary system As–Se. But in proximity to the compound $GeSe_2$, stable glasses are obtained neither in the system Ge–Se nor in the ternary system.

In 1968 we, together with Yegorova (Kislitskaya), studied glass-forming and properties in the system Sb–Ge–Se. At the same time this system was being studied by Borisova and Pasin,[87] Haisty and Krebs,[88] and Brau and others.[89] The boundaries of the glass-forming region — much smaller here than that in the system As–Ge–Se — are determined by the absence of glasses in the binary system Sb–Se and, adjacent to it, compositions with low germanium content. On the section close to $GeSe_2$ the glass-forming boundaries essentially coincide.

Certain problems appeared at the investigation of the glass-forming beyond the pseudobinary cut $GeSe_2$–$Sb_2 Se_3$: when quenching these glasses from 700°C to room temperature boiling-up of the melt with the formation of crystalline drops in the upper, cooler part of an ampul was observed. The range of these compositions is confined in the triangle $GeSe$–$Sb_2 Se_3$–$GeSe_2$, and the condensate amount at that increases sequentially as the cut $Sb_2 Se_3$–$GeSe$ is approached. X-ray structural analysis showed that the condensate is largely composed of germanium monoselenide. The initial composition isn't, apparently, identical to the composition of the produced vitreous melt. Hence it is erroneous to rank this region with the glass-forming region in the system Sb–Ge–Se. We should notice that at 5 g of a substance quenching, GeSe emission is slight and can go unnoticed, which, obviously, took place in the above-cited works. Linke,[90] referring to our data, notes that the emission of GeSe can be reduced to a large extent through the lowering of the quenching temperature to 500°C. We had established this fact as long ago as in 1964, while investigating the system As–Ge–Sn(Pb)–Se, where a similar phenomenon was observed, but in such cases vitreous melts appeared to be partially crystallized. The same takes place in the system Sb–Ge–Se.

However, in the mid-1970s Pasin and Orlova, while investigating crystallizability of the glasses in this system by the method of differential-thermal analysis, had revealed several noncrystallizing glasses within the narrow composition limits beyond the line $GeSe_2–Sb_2Se_3$ with more than 20 at.% of germanium and up to 20 at.% of antimony.[91] The properties of these glasses and the causes for their formation are described in detail below. On the line $GeSe_2–Sb_2Se_3$ glasses can be formed only if both components are taken in the proportion 1:2. Neither in the region of compositions, enriched with stibium selenide, nor in the region with the larger content of germanium selenide are glasses formed.

Bismuth enters into glasses in the amount up to 5 at.%, i.e., at the proportion Ge/Bi = 4:1 and more. A somewhat larger glass-forming region is presented by Pasin and Borisova,[92] where the minimum bismuth content is 10 at.%. Such difference is attributable to the application by the authors of the more severe melt quenching — from 850°C to room temperature.

Glasses with bismuth are easily crystallized and, according to the evidence from the roentgeno-phase analysis, obtained by Pasin and others,[93] Bi_2Se_3 and $GeSe_2$ are emitted at crystallization. Glass-forming regions revealed in the systems As–Ge–Se, Sb–Ge–Se, and Bi–Ge–Se, due to the investigations performed by different authors, are presented in Figure 7. One can see that, according to our data, the glasses are formed in a smaller region, but this is determined by the synthesis regimes. We evaluated the boundaries of the glass-forming region at the composition change of 2.5 at.% of each component in the conditions of the stable glass formation.

1.4.3.2 Properties and Structure of the Glasses in the Systems As–Ge–Se, Sb–Ge–Se, Bi–Ge–Se

The changes of the glass properties in the system As–Ge–Se are demonstrated in Figures 8 and 9 as the surfaces obtained through the lines of equal value properties and also as the diagram's "composition-property", plotted for the binary glasses and various cuts of the ternary system (Figures 10 to 15). It is seen from the pictures that in the seemingly elementary system, consisting of only three types of atoms, the properties of glasses do not change monotonically. In the majority of cuts threefold fractures on the curves are observed. It comes especially noticeable in Figures 11 to 14, where one can see the values of microhardness, the glass-transition temperature, and dilatometrical softening temperature by isogermanium cuts.

Further on we can see that the peculiar points on the property curves coincide not only with certain chemical compounds and eutectics. On the diagram of phase equilibrium, according to the data obtained by Vinogradova,[7] one can see two binary and three ternary eutectics and only one ternary compound. It should be noticed that the phase diagram represents one of the physicochemical characteristics of the system, i.e., the position of the boundaries of the crystallization fields and the nature of the crystalline phases are apparently connected with the structural changes in this

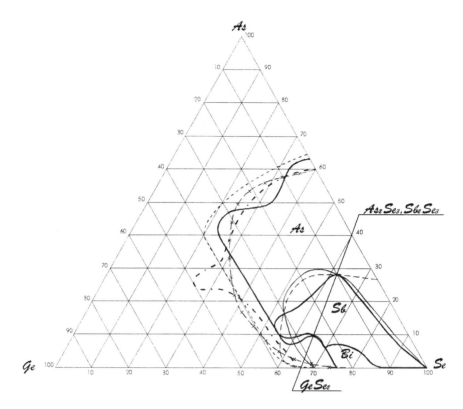

FIGURE 7. Glass-forming regions in the systems As–Ge–Se, Sb–Ge–Se, and Bi–Ge–Se. (———) our results; (--------) Savage's data;[84] (— · — · —) Baydakov's data;[83] (— × — × —) Goryunova and others' data.[26]

system. What structural changes can take place in Ge–As–Se glasses? To answer this question it's necessary to have an idea of coordinating atomic numbers in the compounds formed during glass-transition.

In 1961 we supposed that germanium in the composition range with an excess of chalcogene formed four bonds. This assumption was most likely because it proceeded from the atomic coordination in crystalline germanium and in the compound $GeSe_2$.

Arsenic in the elementary state and in the compound As_2Se_3 is of a three-bond type. It was logical to suppose that the coordinating arsenic number, equal to three, is retained in glass. The two-bond type of chalcogenes in the majority of compounds and also in glasses was already beyond question in the 1950s. It is worth noting that in the following years these assumptions gained full verification in the series of works where the glass structure As–Ge–Se had been investigated by various physical methods (2.4.11).[8]

Proceeding from this, we examine the processes taking place on the synthesis of glasses in the binary systems As–Ge and Ge–Se. Adding arsenic to selenium gives the reaction which in general form may be written as follows:

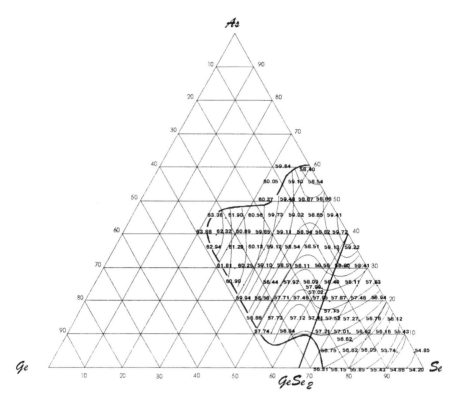

FIGURE 8. Surface of volume atomic concentrations ($N \times 10^3$) of the glasses from the system As–Ge–Se.

$$As_p + Se_n = p\left(AsSe_{3/2}\right) + \left(n - 1.5p\right)\left(Se_{2/2}\right)$$

where n and p are the indices which symbolize content of the given element in glass in atomic percentage. Here and later on, marking, as Myuller does, structural units by fractional indices, we emphasize the polymer character of the structure. It is obvious that when the structural units of $AsSe_{3/2}$ appear between selenium chains, the crystallizability must decrease, which we observe by the experiment. It is not difficult to tally up that the changeover from the chains with dozens of atoms where the selenium structure prevails, to the chains with the units of selenium atoms where the elements $AsSe_{3/2}$ and selenium chains must equally influence the properties of glasses, takes its place adjacent to the composition $As_{20}Se_{80}$. Actually, here exists the eutectic between selenium and As_2Se_3.

In glass with the composition $As_{25}Se_{75}$ (n = 3p) arsenic atoms are connected through two selenium atoms, i.e., here are formed structural units ›As–Se–Se–As‹.

Feltz, who describes in detail the structure of arsenic-selenium glasses, among other things intimates that precisely near this composition change occurs in the

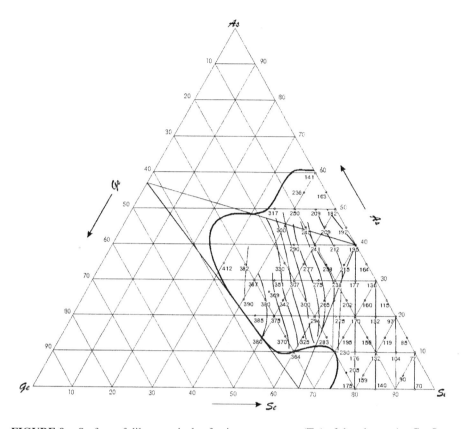

FIGURE 9. Surface of dilatometrical softening temperature (T_{ds}) of the glasses As–Ge–Se.

course of such properties as dielectric penatrability and molar polarization (3.2.1.5).[8] However, in Figures 11 and 12 it is seen that on the curves of microhardness and T_g the first fractures are observed near the composition $As_{35}Se_{65}$ (n = 1.85p), where a buildup of structural units ›As–Se–As‹ occurs, and there remain only 50% of structure correlating with the composition $As_{25}Se_{75}$. Furthermore, all the way to the compound As_2Se_3 (100% of structures $AsSe_{3/2}$ n = 1.5p) the properties change gradually, because only the further buildup of structural elements $AsSe_{3/2}$ occurs. Arsenic selenide is also formed in vitreous state because its structure is polymeric and represents hinge-correlated trigonal pyramids. However, glass-forming doesn't end here. On further addition of arsenic the structure is formed, which accords with the compound AsSe, where the bonds As–As appear, namely,

$$
\begin{array}{cc}
\text{-Se} & \text{Se-} \\
& \rangle \text{As-As} \langle \\
\text{-Se} & \text{Se-}
\end{array}
$$

(Str. 3)

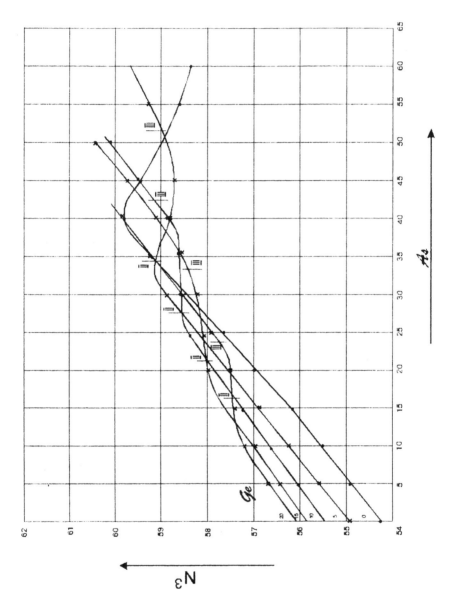

FIGURE 10. Alteration of volume atomic concentrations ($N \times 10^3$) of the glasses As–Ge–Se along the cuts Ge-const.

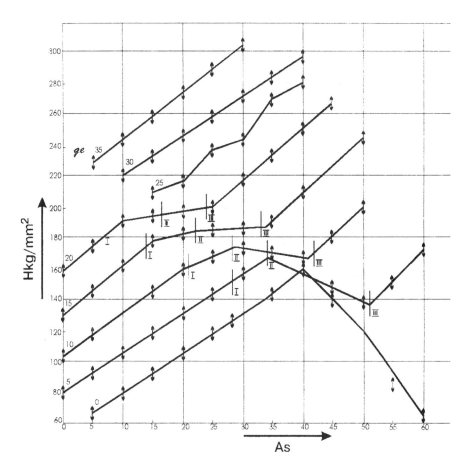

FIGURE 11.   Microhardness (H) alteration in the glasses As–Ge–Se along the cuts Ge-const.

Each arsenic atom holds two bonds with selenium which assures hinge correlation of polymer structure.

Glass-forming comes close in proximity to the composition $As_{60}Se_{40}$, where there is 50% of nonglass-forming structure As_2Se in which arsenic is connected with selenium only through one of three bonds and, as Myuller thinks, it forms closed rings of the following type:

$$
\begin{array}{ccc}
 & As \;—\; As & \\
 \diagup\;\diagdown\;\diagup\;\diagdown & & \\
As \;—\; Se\; Se\; —\; As & & \\
 \diagdown\;\;\;\;\diagup & & \\
 & Se & \\
 \diagup\;\diagdown & & \\
As \;—\; As & &
\end{array}
$$

(Str. 4)

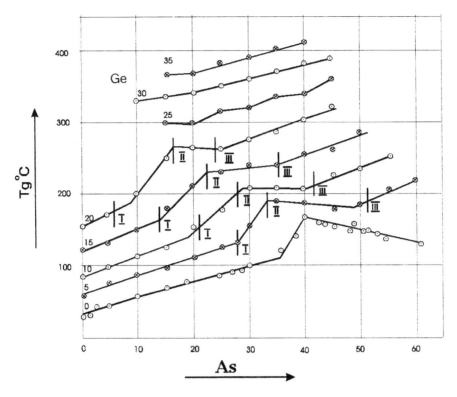

FIGURE 12. Glass transition temperature change ($T_g = 10^{13}$ P) in the glasses As–Ge–Se along the cuts Ge-const. The data by Nemilov.[78]

Such assumptions about the structure of glasses with excess of arsenic as compared to the compound As$_2$Se$_3$ are experimentally confirmed by the drastic decrease in the property values in this compositions range.

Consider the binary system Ge–Se. Germanium added to selenium apparently gives the reaction:

$$Ge_m + Se_n = m\left(GeSe_{4/2}\right) + (n - 2m)\left(Se_{2/2}\right)$$

Stable glasses are formed in this system up to the composition Ge$_{20}$Se$_{80}$, the structure of which can be presented as the coupling of two germanium-selenium tetrahedrons with the bridge of two selenium atoms, namely:

$$
\begin{array}{ccccccc}
 & | & & & & | & \\
- & Ge & - & Se - Se & - & Ge & - \\
 & | & & & & | & \\
\end{array}
$$

(Str. 5)

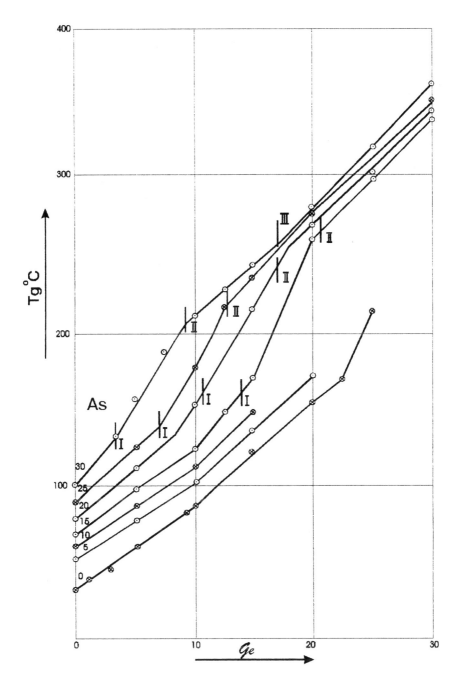

FIGURE 13.   Alteration of T_g of the glasses As–Ge–Se along the cuts As-const.[78]

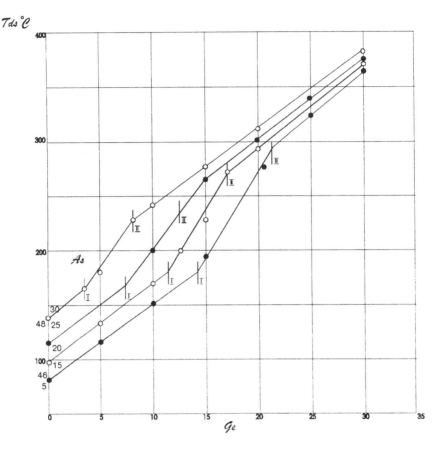

FIGURE 14. Alteration of T_{ds} of the glasses As–Ge–Se along the cuts As-const.

Malaurent and Dixmier,[94] who studied the structure GeSe$_4$ by the KRR method, describe it the same. Some scientists treat these structures as the combination of clusters GeSe$_{4/2}$, however, at accumulation of such structural units crystalline GeSe$_2$ is bound to occur. Crystalline structure GeSe$_2$ is similar to α-GeS$_2$, where distorted tetrahedral structural groups are connected through sulfur atoms — half in vertices, half in ribs. Such a structure as well as SiS$_2$ can't form stable glasses, hence there is no glass-forming near this compound.

The composition Ge$_{25}$Se$_{75}$ itself (n = 3m), where the quantity of structural elements GeSe$_{4/2}$ ranges up to 50%, is produced in the glass form only in the quenching regime. Otherwise crystallization occurs and the initial crystalline phase represents GeSe$_2$. Withfurther increase of germanium, crystallization is bound to occur and germanium diselenide is always evolved as a crystalline phase. The transition from chains with dozens of atoms to chains with per-units of selenium atoms takes place in the proximity of the composition Ge$_8$Se$_{92}$, which is also eutectic. Therewith the curve "composition-T$_g$" (Figure 13) shows a slight bend. By analogy with the system

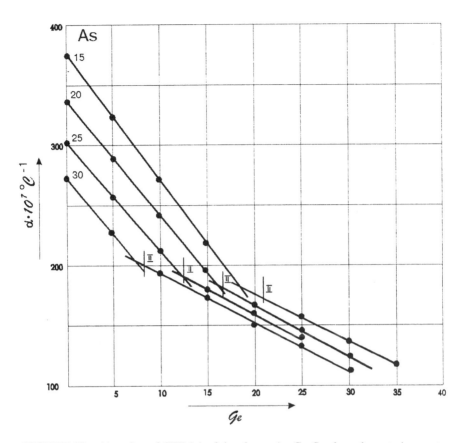

FIGURE 15. Alteration of CTE (α) of the glasses As–Ge–Se along the cuts As-const.

As–Se, the second change of properties in the system Ge–Se should be expected in proximity to the composition $Ge_{25}Se_{75}$.

Let us examine ternary systems As–Ge–Se and Sb–Ge–Se. It would appear natural that the absence of a noticeable heterodynamism and with only one bond type at hand, all the elements in the three-component system have coordination numbers and valences that are typical of them in coinciding binary systems. If so, then in the system As–Ge–Se we can write the following reaction for the region with excess of chalcogene:

$$Ge_m + As_p + Se_n = m\left(GeSe_{4/2}\right) + p\left(AsSe_{3/2}\right) + \left[n - \left(2m + 1.5p\right)\right]\left(Se_{2/2}\right)$$

The first structural measurements here should, apparently, take place in the compositional region where selenium chains give way to the three-dimensional structures with hinge-like bonds interspersed with one and two selenium atoms (with $n = 3m + 1.85p$). Then in the region of compositions $3m + 1.85p \geq n \geq 2m + 1.5p$, i.e., up to pseudobinary cut As_2Se_3–$GeSe_2$, further increasing occurs of the total covalence correlation of the system with the gradual replacement of articulated bonds of two selenium atoms with structural units $GeSe_{4/2}$ and $AsSe_{3/2}$, correlated through one selenium. The fact engages our attention that in the system Ge–Se after the composition $Ge_{25}Se_{75}$ there are no glasses, but on the addition of arsenic (no less than 10 at.%) the glass-forming takes place again. We conclude that the mixed structures of

$$\begin{matrix} \backslash & & & & / \\ As & - & Se & - & Ge & - \\ / & & & & \backslash \end{matrix}$$

(Str. 6)

type prevail in the system As–Ge–Se. Elementary calculation demonstrates that arsenic complex must be located at least after two germanium ones. Such interpenetration of trigonal and tetrahedral structures provides no way for the jointing of germanium-selenium tetrahedrons along the ribs and as a result hinge-type joints appear in the system. Here, as well as in the system Ge–Se the presence of the associated $GeSe_2$ groupings would have resulted in its occurrence in the form of a crystalline phase.

Pseudobinary cut $AsSe_{3/2}$–$GeSe_{4/2}$ (n = 2m + 1.5p) is followed by the glass-forming region with the shortage of chalcogene for saturation of germanium and arsenic valence bonds (Figure 7).

To describe the processes taking place in this region, we use Myuller's thermo-chemical calculations.[6] According to these data, enthalpy change on germanium reaction with selenium yields $GeSe_{4/2}\Delta H^{\circ}_{298} = 68 \frac{kcal}{mol}$ and $AsSe_{3/2} = 54 \frac{kcal}{mol}$ (with sulfur correspondingly $\Delta H^{\circ}_{298} = 100$ and $66 \frac{kcal}{mol}$. Consequently, germanium is the first to enter into the reaction with chalcogene and arsenic excess valences become saturated while yielding the bonds As–As as in the system As–Se.

With further discussion it's necessary to introduce the limiting condition. In view of the fact that in the system As–Se glass-forming comes to the end in proximity to the composition $As_{60}Se_{40}$ (n = 0.65p), we apparently should suppose that in glasses of the ternary system there also can't be bonds As–As in greater quantity, because devitrification would be inevitable.

However, glass-forming in the ternary system isn't over here. The concepts exist that at further exhaustion of the system in chalcogene the structural nodes $AsAs_{3/3}$ and even $GeGe_{4/4}$ are formed.

We consider it obvious that the presence of structures typical of elementary arsenic and germanium in the region where they represent up to three quarters of glass composition rules out the glass-forming possibility. The occurrence of glasses with such compositions is explained by the formation of compounds AsGe and As_2Ge. The structure As_2Ge, described in detail by Bryden,[95] comprises the column of curved pentagons, consisting of two germanium and three arsenic atoms, coupled in layers through the fourth arsenic atom. One can envision that glasses are formed when chalcogene atoms appear in the structure, and they divide the columns of germanium-arsenic pentagons up by hinged joints into the isolated "packets" (Figure 16). As the calculation shows, glass-forming comes to the end when there are more than four pentagons in the "packet". It's obvious that if these considerations are true, the third change in the structure must occur at n = 0.65p + 2m.

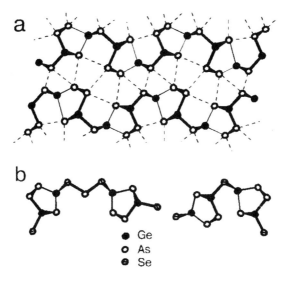

FIGURE 16. (a) View of a layer of a crystalline network GeAs$_2$ (the fourth bond Ge and the third bond As are not shown); (b) possible variants of pentagons Ge$_2$As$_3$ replacement in the glass network.

Hence in the system As–Ge–Se there exist four regions of different structures.

1. The region of selenium chain structures
   Compositions: (100 ≥ n ≥ 3m + 1.85p); bonds Ge–Se, As–Se, –Se–Se–Se–
2. The region of spatial structures of germanium and arsenic selenides
   Compositions: (3m + 1.85p ≥ n ≥ 2m + 1.5p); bonds: Ge–Se, As–Se, Se–Se
3. The region of stratified structures
   Compositions: (2m + 1.5p ≥ n ≥ 2m + 0.65p); bonds: Ge–Se, As–Se, As–As
4. The region of germanium-arsenic pentagons
   Compositions: (n ≤ 2m + 0.65p); bonds: Ge–Se, As–Se, As–As, Ge–As

With Ge–Ge bonds appearing glass-forming comes to the end.

According to our views, bendings and fractures on the curves of properties must take place close to the boundary between structural regions. We performed calculations of the edge compositions for different systems' cuts according to the following scheme:

$$n + m + p = 100$$

$$I\ n = 1.85p + 3m;\quad II\ n = 1.5p + 2m;\quad III\ n = 0.65p + 2m$$

E.g., m = 5. Then:

    I. $95 - p = 1.85p + 15$ $p = 80/2.85 = 28$
    II. $95 - p = 1.5p + 10$ $p = 85/2.5 = 34$
    III. $95 - p = 0.65p + 10$ $p = 85/1.65 = 51$

On Figures 11 to 15 the position of the calculated extreme points is indicated by vertical lines. One can see that the discrepancy between calculation and experiment lies, with rare exceptions, within each component change of 1 at.%. Considering the errors which may take place at charge preparation (each point is obtained in calculations on the samples of various meltings), differences in methods of properties' determination, and relatively low accuracy of calculations, such agreement can be considered demonstrative enough. Hence, any nonmonotonous changes of glass property occur as a result of structural changes, connected with other structural units appearing, i.e., of new structural bonds in the end.

We have examined the system As–Ge–Se so extensively first and foremost because it involves the largest glass-forming region which contains all types of structures possible in glasses, namely: chain, stratified, and three-dimensional ones. In this connection the greatest change of properties is observed here: the glass-transition temperature and dilatometrical softening temperature undergo a change from 37 to 60°C, with pure selenium to ≈400°C in the region of compositions enriched with germanium and arsenic: α — from 40 to 10×10^{-6}°C^{-1}; microhardness — from 55 to ~300 kg/mm^2. According to this, the system As–Ge–Se used to be taken as the principal one at production of complex glasses.

Compositional changes of properties in the system Sb–Ge–Se are demonstrated in Figures 17 and 18 as the surface of concentration of atoms and diagrams constructed by the cuts Sb — 10 at.% and Sb — 15 at.% in comparison to the properties of glasses As–Ge–Se. The agreement between the values of T_{ds} and α and also analogy in the course of microhardness show that antimony forms in glass network structural elements SbSe$_{3/2}$ similar to AsSe$_{3/2}$. Structural changes manifest themselves especially clear in the dependence of atoms' concentration on the composition (Figure 17). In the systems with antimony as well as in the system As–Ge–Se a rise of atomic concentration occurs as a result of the replacement of chain selenium with spatial structures SbSe$_{3/2}$ and GeSe$_{4/2}$. Maximum values of [N] are observed in glasses, which are on the bisectrix of the angle GeSe$_2$–Sb$_2$Se$_3$–GeSe$_4$: in their structure the bonds between atoms of metals are performed by monatomic and diatomic selenium bridges.

The same concentration values take place in proximity to pseudobinary cut Sb$_2$Se$_3$–GeSe$_2$ in the region with a predominance of more dense antimony structures. As the number of "friable" structural units GeSe$_{4/2}$ increases, the total atomic concentration decreases. Absolute values [N] in the system Sb–Ge–Se are lower than in As–Ge–Se, which derives from the larger size of the antimony atom. Microhardness of antimonous glasses is also slightly lower in its absolute value than in arsenic ones.

It will be recalled that microhardness is measured by the method of pressing the diamond pyramid in and, as a result, the distraction of the surface layer of glass occurs, i.e., the rupture of chemical bonds. That is why microhardness degree is determined, all other factors being equal, by the number of broken bonds and their

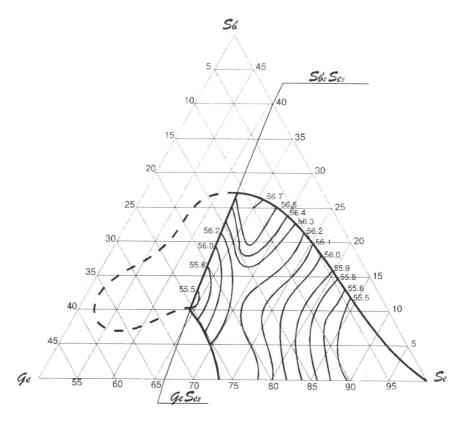

FIGURE 17. Glass-forming region and concentration surface of the particles $N \times 10^3$ in the system Sb–Ge–Se.

energy. As the energy quantity of the bonds As–Se and Sb–Se is approximately identical, the difference in microhardness should be, apparently, connected with the difference in volume concentration of particles. Microhardness values measured by us satisfactorily conform with the data by Pasin and Borisova.[96] We measured T_{ds} and α only for glasses of two cuts with the constant content of antimony equal to 10 and 15 at.% (Figure 18). Savage and others studied more closely the dependence of linear expansion of glasses Sb–Ge–Se on the composition.[97] It is demonstrated that, according to our data as well, α changes in the range from 13 to 28×10^{-6}°C^{-1}. Especially noteworthy is the availability of glasses along the cut GeSe$_2$–Sb$_2$Se$_3$. Glass-forming here occurs only in the middle part, where no one chalcogenide is in a prevailing amount. This is one more experimental verification of the nonglass-forming character of the structures GeSe$_2$ and Sb$_2$Se$_3$. It seems quite obvious that glasses are formed only as a result of the appearance of heterostructures of the type:

$$
\begin{array}{ccc}
\diagdown & & \diagup \\
-\, \mathrm{Ge} \, - & \mathrm{Se} \, - & \mathrm{Sb} \\
\diagup & & \diagdown
\end{array}
$$

(Str. 7)

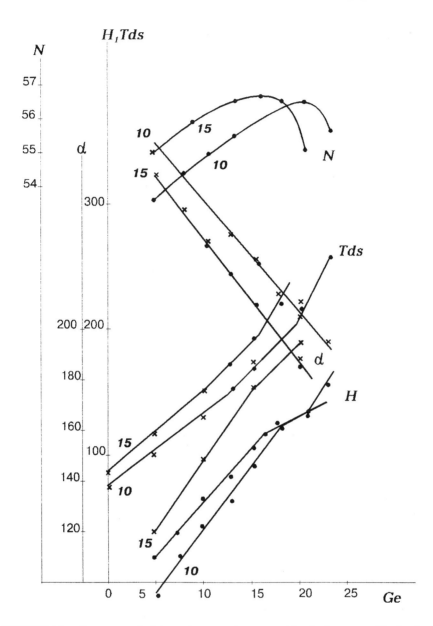

FIGURE 18. Glass properties' dependence on the composition in the system Sb–Ge–Se in comparison with the glasses As–Ge–Se (Sb–As-const.). (•••) Sb–Ge–Se; (×××) As–Ge–Se.

and this itself is the testimony in favor of the model of chemically ordered bond network. Rather insufficient glass-forming in the region of compositions with the lack of selenium for the formation of the complexes $GeSe_{4/2}$ and $SbSe_{3/2}$ is explained by the absence of the compounds SbSe and GeSb (in contrast to AsSe, GeAs, and $GeAs_2$), i.e., by impossibility to form the bonds Ge–Sb. The availability of several glasses in this region is determined by the ternary compound $Ge_4Sb_2Se_4$, found by Orlova and others during the investigation of the phase diagrams of the subsystems Sb_2Se_3–$GeSe_2$ and Sb_2Se_3–$GeSe$.[98] The conclusions about the structure of glasses in the system Sb–Ge–Se, based on our investigations of glass-forming and the above-listed properties, were used by Borisova and Pasin to explain the results of the studies of electric conductivity,[96] and also by Korepanova and others, who studied thermal expansion and ultrasound velocity.[99] Gutenev and others, while studying dielectric and magnetic properties of glasses in the system Sb–Ge–Se, established the systematic increase of IR component of electric susceptibility with the enhancement of antimony content and the negative deviation from the linear course of dependence of Van-Flick's polarization paramagnetism for glasses of pseudobinary cut. The authors explain these facts by statistically steady distribution of the structural elements $SbSe_{3/2}$ and $GeSe_{4/2}$ in the glass network.[100]

1.4.3.3   Glasses Based on the Systems As–Ge–Se and Sb–Ge–Se, Containing Heavy Elements of the IV Group — Tin and Lead

The influence of tin and lead investigation upon glass-forming and properties of glasses is of both theoretical and practical interest. Crystalline structures of these elements' sulfides and selenides aren't glass-forming. SnS and SnSe are crystallized in the distorted lattice of sodium chloride, isomorphically with GeS and GeSe.[101] Lead tends even more to the conservation of the unshared S-electron pair and, as a result, it performs only one compound with selenium PbSe (for tin dyselenide is also known) which is crystallized, according to the data by Krebs, in the lattice NaCl at the sacrifice of six-bond formation on p-orbitals each way from the nucleus of the Pb atom.[102] This, apparently, can explain the poor attempts to introduce tin and lead into glasses as chalcogenides or to create vitreous oxygen-free systems on their basis.

In the system As–Sn–Se the glass-forming region isn't large: tin can be introduced into the arsenic-selenium glasses in the amount not exceeding 8 at.%. In the system Ge–Sn–Se the glass-forming region is also insignificant and is flanked on the line of compositions Ge–Se. The maximum content of tin is equal to 12 at.%. The attempts to form glasses in the ternary systems with lead (As–Pb–Se, Ge–Pb–Se) have failed. Either complete crystallization or formation of distinctly diphase melts is observed.

The glass-forming regions with sufficiently stable glasses were revealed in the four-component systems obtained on germanium replacement with tin and lead in the system As–Ge–Se, where their maximum content is equal to 15 at.%. The systems As–Ge–Sn–Se, As–Ge–Pb–Se, and As–Ge–Sn–Pb–Se were studied in detail by Yefimov. He observed in them considerable decrease of glass formation, as the content of heavy element enhances in comparison to the initial system, which is clearly shown in Figures 19 and 20, takes place predominantly in the compositional regions mostly rich in arsenic. Violation of glass-forming is of a special character in that in the process of glass cooling after synthesis from temperatures higher than 600°C, there occurs vigorous dissociation of melt with expansion of free arsenic that is condensed in the form of microcrystalline generations on the walls of an ampul. Chemical constitution of the condensate is established by spectral and roentgenostructural analyses. Apparently the amount of germanium which remains after its replacement with heavy elements isn't enough to connect up all arsenic and the bonds Pb–As and Sn–As can't be formed in glasses.

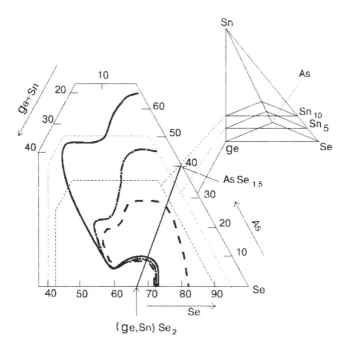

FIGURE 19.   Glass-forming regions in the system As–Ge–Se–Sn (5 and 10 at.% Sn).

A striking likeness between the process of arsenic "exclusion" by tin and lead and deriving from it the essentially equal position of glass-forming boundaries in the systems As–Ge–Sn–Se and As–Ge–Pb–Se is indicative of the likeness of structural complexes formed by atoms of tin and germanium in glasses with the lack of selenium.

The most evident individual difference between the systems with tin and lead is observed in proximity to the compositions from pure selenium to the pseudobinary cut $AsSe_{3/2}$–$GeSe_{4/2}$. Insertion of tin only slightly changes the glass-forming region of the initial system, whereas in the lead glasses glass-forming considerably decreases. In addition, between crystalline melts and monophase glasses a considerable region of diphase substances is observed. The transition from completely crystallized compositions to the vitreous ones is emanating gradually at that. As the content of germanium increases at the expense of selenium, the recent surfaces of chips — tarnished and close-grained in crystalline specimens — take on the luster typical of glass and the specific shell structure. However, a great number of almost imperceptible inclusions are clearly visible on these surfaces when viewed through a microscope of 500 power in reflected light, and their dimensions are decreasing

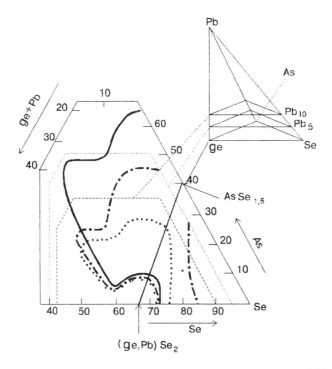

FIGURE 20. Glass-forming regions in the system As–Ge–Se–Pb (5 and 10 at.% Pb).

gradually. The same inclusions are viewed, few in number, of superficially vitreous specimens of compositions of ternary systems Ge–Pb–Se and As–Pb–Se. The X-ray structural analysis of such manifestly diphase specimens didn't show the crystalline phase presence in them.

We managed to establish the crystalline character of insertions and their compositions through the roentgenophase analysis of those specimens where at repeated heat treatment the stratification into two phases is observed. The crystalline phase therewith forms the lower layer, the vitreous phase the upper one. Viewing of the upper layer through the microscope had revealed the presence of monophase glass; roentgenophase analysis of the lower one determinated it as lead selenide. These results are in accordance with the data obtained by Shkol'nikov and Borisova, who settled PbSe elimination at the crystallization of vitreous materials in the system $AsSe_{1.5}$–Pb, with lead content overcoming 5 at.%.[103] This demonstrates again the imperfection of roentgenophase analysis as the way of detecting partial glass crystallization.

In conclusion, it should be noted that the elimination of microcrystalline insertings of PbSe in lead-containing glasses revealed by us gives grounds to the evaluation of the results of researches carried out by other investigators in a somewhat different way. For example, the unsuccessful attempt to crystallize out the natural "glass", revoredit, undertaken by Petz and others,[104] confirms, in our opinion, the fact that revoredit is a diphase system where lead is presented in the form of small crystals of PbS, allocated in mass of noncrystallizing glass As_2S_3. The reduced transparency of the glasses $PbS–As_2S_3$ in the IR region was even established by Frerichs.[11] It can be explained only by the presence of the microcrystalline phase PbS. Note that the diphase character of lead glasses had enabled Yefimov to work out the IR glass filter with the moving absorption edge.

The changes in the character of glass property dependences from the composition on the insertion of tin are demonstrated in Figure 21 in the form of surfaces of atomic concentrations, and in Figure 22, where one can see the change in glass-transition temperature, microhardness, and atomic concentration along the isoarsenic cut As = 15 at.%.

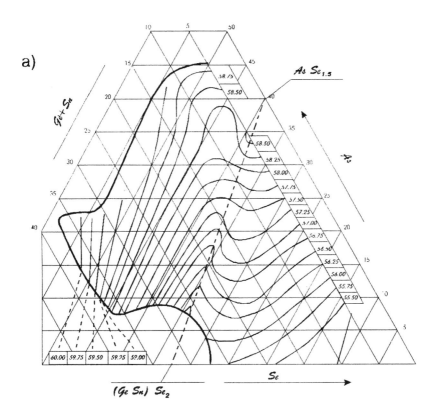

FIGURE 21. Concentration surfaces (a) in the system $As–Ge–Sn_5–Se$; (b) in the system $As–Ge–Pb_5–Se$.

As is evident from the pictures, in the region of selenium chains and spatial structures, i.e., up to n = 2m + 1.5p, germanium replacement with tin doesn't change properties of glasses of the principal system and on this basis we can come to the conclusion about their structural likeness. Differences in absolute values of concentrations as well as in glasses with antimony derive from the larger sizes of tin atoms.

On transition to the region with the lack of selenium in tin glasses, the concentration of atoms abruptly increases and at the same time viscosity and T_{ds} decrease,

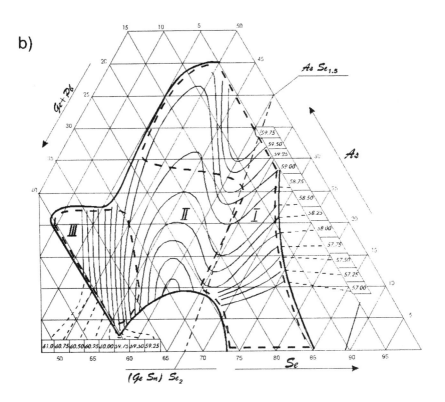

b)

FIGURE 21 (Continued)

with no changes of α. This points to the fact that spatial complexes of the principal system are conserved, but with structural elements of another type. Yefimov offers for this region the structural elements $SnSe_{2/2}Se_{1/3}$. In such a structure an atom of tin is located in the vertex of the trigonal pyramid and produces three bonds, as in the crystal SnSe, but as there are no other atoms of tin in the closest neighborhood and under the influence of germanium-arsenic vitreous network two atoms of selenium form only two bonds, respectively, i.e., the hinge-like bonds appear in the structure. The structural formula of such complexes is represented in Figure 23. This model helps to explain the behavior of tin in oxygen-free glasses in compositions with a lack of selenium. In the region with selenium excess, i.e., up to the line As_2Se_3–(Ge, Sn) Se_2, tin behaves like germanium.

Taking into account uniformity of crystalline structures of tin and lead selenides we can suppose that in glasses they also form structural elements of one and the same type, i.e., being under the influence of a spatial vitreous network, lead forms in monophase glasses the elements $PbSe_{2/2}Se_{1/3}$. This is confirmed by the agreement between the values of atomic concentration and microhardness in lead and tin glasses in the region, with considerable lack of selenium and complete resemblance of glass-forming regions at an equal content of elements. The maximum content of tin and lead is equal to 15 at.%. Therewith in each of four-component systems only one

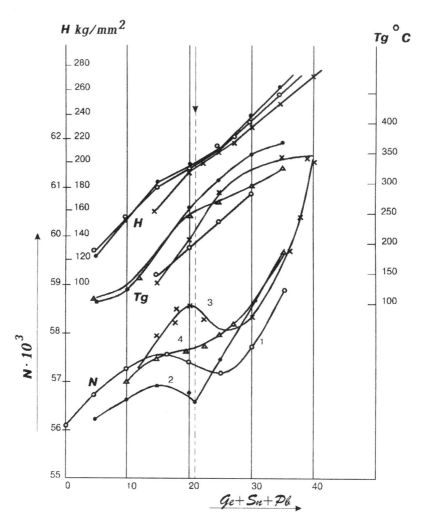

FIGURE 22. Properties of glasses in the systems containing lead and tin in comparison with the system As–Ge–Se. (1) As_{15}–Ge–Se; (2) –As_{15}–Ge–Sn_5–Se; (3) As_{15}–Ge–Pb_5–Se; (4) As_{15}–Ge–$Sn_{2.5}$–$Pb_{2.5}$–Se.

glass is formed (with components' replacement at every 5 at.%). In the system with tin the composition of this glass is $As_{10}Ge_{15}Sn_{15}Se_{60}$; in the system with lead — $As_5Ge_{25}Pb_{15}Ge_{55}$.

In the glasses of a quinary system As–Ge–Sn–Pb–Se the character of property dependence on the composition is determined by summation of partial values introduced by tin and lead.

The synthesis of melts in the system Sb–Ge–Sn–Se was carried out in time-temperature regimes similar to those in the system Sb–Ge–Se. The glass-forming regions in the systems with the content of tin equal to 5, 7.5, and 10 at.% are demonstrated in Figure 24. As well as in other systems containing heavy elements

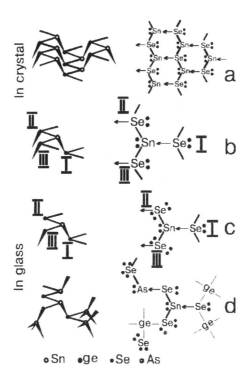

FIGURE 23. (a) View of a layer of the crystalline network SnSe; (b) structural element of the network SnSe ($SnSe_{3/3}$); (c) structural element $SnSe_{1/3}Se_{2/2}$; (d) example of possible variants of the replacement of the elements $SnSe_{1/3}Se_{2/2}$ in the glass network.

of the IV group, the glass-forming decreases when the quantity of tin enhances, therewith the glass-forming region reduces at the expense of sections with low content of germanium. In the section adjacent to $GeSe_2$ the glass-forming edge doesn't practically change. Stable glasses are also formed in the region with the lack of selenium, especially when 7.5 at.% of tin is introduced.

The formation of stable glasses in the region lacking selenium is evident of the fact that here, as well as in the system As–Ge–Sn–Se, valence condition of tin changes, and in spite of tetrahedral structures $SnSe_{4/2}$, trigonal ones $SnSe_{2/2}Se_{1/3}$ are formed, with two bridge bonds and one donor-acceptor bond in selenium. Inserting 15 at.% of tin causes crystallization of all melts with expansion of $SnSe_2$ as the first crystalline phase.

The above-given view of glass structures is verified by the results of investigations of properties' changes with composition.

In Figure 25 one can see the change of the total atomic concentration coming from germanium and tin content in glasses with the constant amount of antimony equal to 5 (I) and 15 at.% (II) in comparison to the system Sb–Ge–Se.

Insertion of larger amounts of tin in place of germanium causes the gradual decrease of atomic concentration because the structural elements $SnSe_{4/2}$ are looser than $GeSe_{4/2}$. Coming to the region with the lack of selenium, atomic concentration

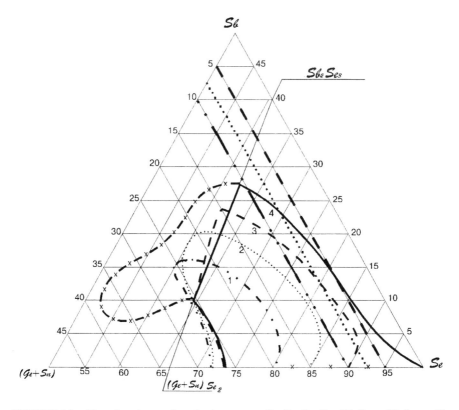

FIGURE 24. Glass-forming regions in the system Sb–Ge–Sn–Se. (1) Sn_{10}; (2) $Sn_{7.5}$; (3) Sn_5; (4) Sn_0.

increases abruptly due to valency state of germanium and tin when the structures of the type $MeSe_{2/2}Se_{1/3}$ are formed. Therewith practically complete independence on the amount of tin is observed (all the curves are merged into one). T_{ds} isn't higher than that of the analogues in the system As–Ge–Se, hence the maximum T_{ds} of glasses in the systems Sb–Ge–Se and Sb–Ge–Sn–Se is equal to 280°C.

1.4.3.4   Glass-Forming in the Systems As–Ge–Se and Sb–Ge–Se after Selenium Replacement with Tellurium and Addition of the Elements of the III Group: Ga, In, Tl

We didn't use tellurium as the principal glass-former because according to the notions developed in Section 1.3, there were no grounds to expect stable glasses to be formed. Our considerations are confirmed by the works by Hilton,[80] where the systems Ge–P–Te, Ge–Sb–Te, Ge–As–Te, and their blends were studied, but the stable glasses were not obtained. Savage also ascertains that glasses in the system Ge–As–Te are unfit for commercial manufacturing.[105] Tellurium, however, may be involved in glasses in the form of additions to selenium in the amount up to 35 at.%.

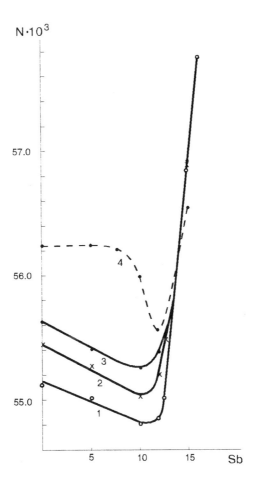

FIGURE 25. Alteration of atomic concentration in the system Sb–Ge–Sn–Se along the cut $(Ge + Sn) = 22.5$ at.%. (1) Sn_{10}; (2) $Sn_{7.5}$; (3) Sn_5; (4) Sn_0.

For the most part the stable glasses are obtained in the system As–Se–Te which had been first studied by Jerger in 1959.[106] Obraztsov and Orlova,[107] who studied this system in detail, found in the subsystem As_2Se_3–Te the eutectics with the content of tellurium equal to 28.6 at.% and established that with tellurium addition T_{ds} of glasses abruptly decreases (from 170 to 60°C). Based on the available data about the systems As–Se–Te and As–Se we had chosen three glass types as the initial ones: $As_{30}Se_{70}$, $As_{40}Se_{60}$, and $As_{50}Se_{50}$, where selenium was being gradually replaced with tellurium. Specimens' weight was 30 g and maximum fusion temperature 900°C. Compositions and some properties are presented in Table 2. It can be seen from the table that, while the content of tellurium is enhancing, density increases linearly, but atomic concentration decreases, i.e., glasses are getting more friable. It is significant that T_{ds} of glasses with the total content of selenium and tellurium remains practically constant and changes only when the amount of arsenic increases. This means that tellurium, like selenium, can form two bonds. According to the data by

Obraztsov and Orlova, all the glasses along the cut $As_{30}(Se + Te)_{70}$ lie in the region of compositions, not crystallizing at the thermal-differential analysis. However, the investigations of crystallization in a wide temperature interval (150 to 550°C) demonstrated that tellurium introduction in place of selenium into the noncrystallizing glass $As_{30}Se_{70}$ causes continuous crystallization, the temperature interval of which enhances as the content of tellurium increases. Noncrystallizing glasses can be formed only on tellurium insertion into the initial composition $As_{50}Se_{50}$, which is apparently explained by the mixture of structures AsSe and AsTe.

TABLE 2
The Properties of As–Se–Te Glasses

Composition in at.%					
As	Se	Te	ρ (g/cm³)	N ×10³	T_{ds}(°C)
30	65	5	4.629	57.73	136
30	60	10	4.695	56.83	138
30	55	15	4.807	56.52	136
30	45	25	4.979	55.33	136
30	40	30	5.052	54.71	132
40	50	10	4.762	57.92	166
40	40	20	4.916	56.46	167
50	30	20	4.925	56.82	182

Hilton and others examined the system Ge–Se–Te and established that with the increase of germanium content T_{ds} changes from 89 to 336°C.[80] Our investigations demonstrated that the tendency to crystallization in these glasses is comparatively high and it enhances with the increase of tellurium and germanium content. All the attempts to reduce the crystallization tendency by adding boron, phosphorus, gallium, and indium were of no success. Tellurium glasses (10 and 20 at.%) were synthesized in the system Ge–As–Se along the cut with an equal content of arsenic. They appeared to be the most stable, but their crystallizability was higher than usual.

On indium and gallium addition to the glasses of the system As–Ge–Se vitreous substances are formed mainly along the line $Ge_{20}Se_{80}$. The maximum amount of additions is equal to 10 at.%. In 1993 the work by Ivanova and others was published, where the authors were examining the system GeS_2–Ga–I.[108] Glassy substances were obtained at ice-cold water quenching 500 to 850°C (depending on the composition). Spectral transmission was studied in the films ~0.9 μm thick. In our view of glass-forming, glasses can't even be expected in this system.

The glass-forming region is much larger with a heavy analogue of these elements — thallium. The maximum amount of thallium in glasses is as large as 30 at.%. The glass-forming region in the system Ge–Se–Tl was studied by Dembovsky,[109] Turkina and others,[110] and Linke and others.[111] Linke showed that the glasses with selenium content ~75 to 87 at.% can be produced both in the form of two vitreous layers and in the form of ingots in the volume of which the drop-shaped insertions of one of the glass phases, i.e., liquation drops, are evenly spread. Turkina, investigating the phase diagram of the glasses Ge–Se–Tl by the methods of thermal-differential and roentgeno-structural analyses, found eight ternary compounds. Only three of them — Tl_2GeSe_4, $TlGeSe_2$, and $TlGe_2Se_3$ — were obtained in vitreous state at dispersion of 5 g and a quenching down to room temperature.

In the author's opinion, the presence of ternary compounds in vitreous state allows one to consider that in the structure of glasses, along with the structural units $GeSe_{4/2}$ and $Se–Se_{2/2}$, there take place ternary structural complexes. Thus, for example, the compound Tl_2GeSe_4 can be presented as tetrahedrons $GeSe_{4/2}$, where two selenium atoms interact with thallium, i.e.,

$$
\begin{array}{c}
Se^- \; Tl^+ \\
| \\
- Se - Ge - Se - \\
| \\
Se^- \; Tl^+
\end{array}
$$

(Str. 8)

According to our data, stable glasses are formed only in the region located between pseudobinary cuts $Tl_2Se–GeSe$ and $Tl_2Se–GeSe_2$. These compositions can be obtained in vitreous state both at melt immediate quenching in the air and at ampules' cooling in a quartz pipe with asbestos filling. It should be noted that in this very part of the diagram there are glass-forming compounds $TlGeSe_2$, $TlGeSe_{2.5}$, and $TlGe_2Se_3$. The compositions located in the triangle $TlSe–GeSe_2–Se$ vitrificate only at a rather tough quenching. All the glassy melts obtained in this region, unlike stable glasses, when viewed under the IR microscope, are either opaque or transparent but partly crystallized out. In the process of crystallization various bubble-like insertions appear in the glass volume. However, in their centers there are no light points typical of bubbles, and on further holding in the temperature range of crystallization the druses of opaque crystalline particles of irregular form are generated from them.

The glass-forming region in the system As–Ge–Se–Tl was first studied by Manvelov in 1986. The investigations were carried out over eight surfaces — sections of concentration tetrahedron with the constant content of arsenic equal to 10, 20, 30, and 40 at.% and of germanium equal to 10, 20, 30, and 35 at.%. Insertion of thallium into the initial system differentially influences the glass-forming. In the region with the lack of selenium, homogeneous glasses are formed on thallium addition — not more than 15 at.%. In the section adjacent to the compound $GeSe_2$, where in the initial system there are no glasses, thallium addition results in glass-forming, its optimal amount therewith equal to 10 at.%. In the region with selenium excess the glasses are formed at thallium additions up to 40 at.%; however, the glass-forming region reduces, gradually moving to the maximum content of selenium. In the system Sb–Ge–Se–Tl, which had been first studied by Kislitskaya and Pavlova, glass-forming decreases even on addition of 5 at.% Tl exactly due to the compositions with a large excess of selenium. At the same time, in the region with lack of selenium where in the initial system the precipitation of crystals GeSe is, as a rule, observed, stable glasses are formed on thallium addition, and their compositions agree to those of the estimated ones. Like in the system As–Ge–Se–Tl, a certain expansion of the glass-forming region toward its second part in the system Ge–Se is observed, established by Feltz and Lyppmann.[112] When the content of thallium enhances up to 10 at.%, the glass-forming near the composition Ge(SeTl)$_2$ increases even more. All the glasses are well transparent when viewed under a IR microscope and sufficiently homogeneous. Further increase of thallium content up to 15 and 20 at.% results in considerable decrease of the glass-forming region which is located mainly along the cut Sb$_2$Se$_3$–Ge(Se,Tl)$_4$ (Figure 26).

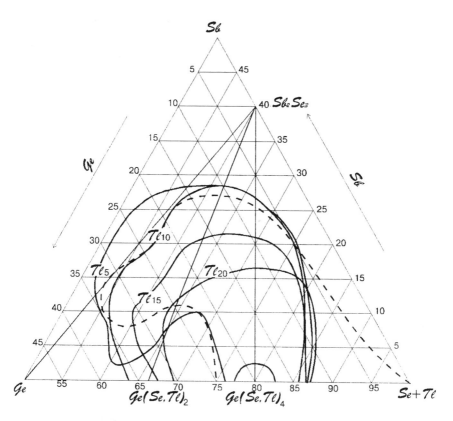

FIGURE 26. Glass-forming regions in the system Sb–Ge–Se–Tl with the content of Tl 5, 10, 15, 20 at.%. (— — —) Sb–Ge–Se; (———) Sb–Ge–Se–Tl.

Complex glasses with heavy elements of IV and V groups of the periodic system and also with tellurium and thallium, where these components represent rather valuable technological additions, were examined at the description of commercial optical glasses' elaboration.

1.4.4 GLASS-FORMING, PROPERTIES, AND STRUCTURAL PECULIARITIES OF SULFUR-CONTAINING GLASSES

The glasses based on sulfides were the first objects in the investigations of oxygen-free systems. It's quite clear that because sulfur is the closest analogue of oxygen, its bonding energy is approximately in 10 kcal/mol higher than that of selenium and tellurium, and besides, in compound with arsenic it forms a widely known natural material obtained in glass form as long ago as the last century — trisulfide arsenic or orpiment. In some works As_2S_3 and As_2Se_3 are examined even as prototypes of ChG, if one bears in mind that in the complex systems their properties don't differ a lot from those in these compounds. Probably, it is correctly reasoned with reference to semiconductor characteristics, because ChG are, in majority, dielectrics with a slight difference between the value of dielectric properties and the character of conductivity; however, physicochemical, optical properties and the structure of this large class of glasses change considerably together with the composition.

In the State Optical Institute sulfur-containing glasses were studied by Ajio.

As was mentioned above, sulfur has a tendency of closed rings forming, and this determines its not very beneficial role in the glass-forming process. Sulfur forms binary glasses only with arsenic; stable glasses at that are formed only in the blends As_2S_3–As_2S_5 with a rather insignificant excursion from stoichiometry. Sulfur insertion into elementary selenium does not cause the expected strengthening of selenium due to the greater strength of sulfide bonds, but the formation of rather easily crystallizing products which are much less hard than the initial material. It is attributable to the appearance of loosely bound sulfur rings in selenium chain structure.

Among the ternary systems with sulfur As–S–Se, Ge–S–Se, and Ge–As–S were inspected. In the system As–S–Se the glass-forming region is rather wide: all compositions along the cut As_2S_3–As_2Se_3 and also in the region with chalcogene excess are vitreous. The absence of glass-forming in the region with the lack of chalcogene derives from the structure of realgar — the compound with the composition of molecular-cyclic constitution: realgar consists of molecules As_4S_4 bound by van der Waals forces.

In the system Ge–S–Se glasses are formed in the small region of compositions. At sulfur content more than 25 at.% the amount of germanium isn't higher than 5 to 10 at.%; at the lower sulfur content it comes up to 20 at.%. The introduction of a greater amount of germanium results in stratification with a crystalline phase isolation.

Even a smaller glass-forming region was found in the system Ge–As–S: the glasses are formed here only on germanium addition to the compositions lying between the compounds As_2S_3 and As_2S_5. The results of glass-forming investigation in binary and ternary systems lead us to the conclusion that glasses are formed mainly in response to the combination of sulfides and it makes sense to insert sulfur into commercial glasses in persistent state in the form of orpiment.

The boundaries of the glass-forming region in the system (As_2S_3)–Ge–Se are almost coincident with those of the system As–Ge–Se. On sulfur addition a certain increase of the glass-forming region along the line of compositions enriched with germanium is observed. Thus, if in the system As–Ge–Se the glass containing 40 at.% Ge can be formed only at a drastic fusion quenching, sulfur containing glass of the same composition is produced at a usual synthesis regime.

Practically complete coincidence of the glass-forming regions demonstrates that sulfur insertion doesn't lead to any considerable structural changes. All compositions along the line (As_2S_3)–Se are vitreous and represent the series of nodes $As(S,Se)_{3/2}$ and selenium chains.

On addition of As_2S_3 to $GeSe_2$ in the amount less than 10 at.% As the glasses are not formed as well as in the system As–Ge–Se. This confirms that mixed structures of the type

$$\begin{array}{ccc} \backslash & & / \\ As & - (S,Se) - & Ge - \\ / & & \backslash \end{array}$$

                                                                              (Str. 9)

are necessary in glass-forming.

Of peculiar interest is the glass-forming along the cut (As_2S_3)–Ge. Addition of 5 and 10 at.% Ge to orpiment results in crystallization of the product, but as germanium content increases, the monophase glasses are formed up to 40 at.% Ge. This phenomenon, paradoxical at first sight, is also attributable to the structural peculiarities of sulfides. Really, on germanium addition to orpiment the following reaction goes: $2As_2S_3 + Ge \rightarrow GeS_2 + 2AsS_2$, i.e., a part of arsenic passes into the realgar nonglass-forming structure. Therewith the crystallization is inevitable. As the amount of germanium increases, all arsenic passes into realgar. After that the bonds As–Ge begin to appear and sulfur enters the mixed structural elements:

$$\begin{array}{ccc} - S & & S - \\ \backslash & & / \\ As & - Ge - S - \\ / & & \backslash \\ - S & & S - \end{array}$$

                                                                              (Str. 10)

As a result, the amount of realgar reduces. The experience of the investigation of the system (As_2S_3)–As–Se shows that glass-forming begins with less than 30% of realgar present in the system. Calculations performed in view of these considerations established that the region of crystallizing compositions along the line (As_2S_3)–Ge lies at germanium content from 3 to 14.5 at.%. Experimental test demonstrated that stable glasses are really formed on the insertion of 2 at.% Ge and also from 15 at.% Ge and more. Absolute values of volume concentration are much higher in sulfide glasses than in pure selenide ones due to the smaller sizes of sulfur atoms.

The influence of heterodynamism introduced by sulfur into the structure of selenide glasses reflects on the properties of sulfide glasses in complex systems, too. Sulfur insertion must seemingly lead to thermal and mechanical strengthening of glasses, whatever the compositions are. Meanwhile, as can be seen in Figure 27, which shows the changes of T_{ds} and T_g in the system (As_2S_3)–Ge–Se as compared to the system As–Ge–Se along the cut As = 20 at.% const., the strengthening influence of sulfur begins to manifest itself only in the region where the bonds Ge–As are predominant.

Fractures on the curves match due to the identity of sulfur and selenium structures. The absence of structural differences is also seen from the values of the α which are practically the same in both systems in the regions with a large germanium content.

The addition of arsenic and selenium in different combinations to orpiment performs the system (As_2S_3)–As–Se. The glass-forming along the cut As_2S_3–AsSe in essence comes to the end here. To replace arsenic with antimony, the glass-forming region markedly reduces. The maximum content of antimony in the system (As_2S_3)–Sb–Se is equal to 15 at.%. In the system (As_2S_3)–As–Se density and microhardness of glasses were measured. Density of atomic packing in sulfoselenide glasses enhances as the content of sulfur increases, but the character of its dependence on the composition doesn't change. At selenium replacement with sulfur, microhardness in glasses with a low content of arsenic is considerably lower than in pure selenium ones, which conforms to the results of the systems' investigations described above. Along the pseudobinary cut As_2S_3–As_2Se_3 microhardness remains constant, because glasses in this case represent the combination of identical elements $AsS_{3/2}$ and $AsSe_{3/2}$. In the region of realgar structures microhardness decreases again, more so the greater amount of sulfur is present in the system. Especially pronounced here is the sulfur tendency to form closed rings which leads to disruptions in polymer structure of selenium glasses.

The glass-forming region with antimony is so small that systematical change of properties is hard to follow, but absolute values of microhardness and T_{ds} are similar to those in arsenic glasses. Hence the replacement of arsenic with antimony may be used at creation of multicomponent glasses for practical application.

Cervinski and others investigated the structure of glasses in the system Ge–Sb–S by EXAFS methods and X-ray scattering.[113] The authors presented a picture of both short- and medium-range ordering in these glasses. However, they had come to the conclusion about the impossibility of describing the structures in terms of a chemically ordered covalent network model, which, in our opinion, is not sufficiently justified.

1.4.5 CHALCOGENIDE GLASSES CONTAINING HALOGENS

To insert the elements of the VII group — namely, iodine, bromine, and chlorine — into ChG, the glasses with specific and valuable properties will be formed. Retaining good transparency for long-wave IR radiations, high chemical stability, and high refractive index, they gain plasticity at room temperature and the nearby transmission edge shifts to the short-wave spectrum range. Such combination of halogen-containing glasses' properties provides new application of ChG, for example, as high-refractable immersion mediums, optical glues, soft IR materials, etc.

The investigation of halogen-containing ChG was begun in the 1960s and the system As–S–I was one of the first to be examined. Thus, Pearson and colleagues

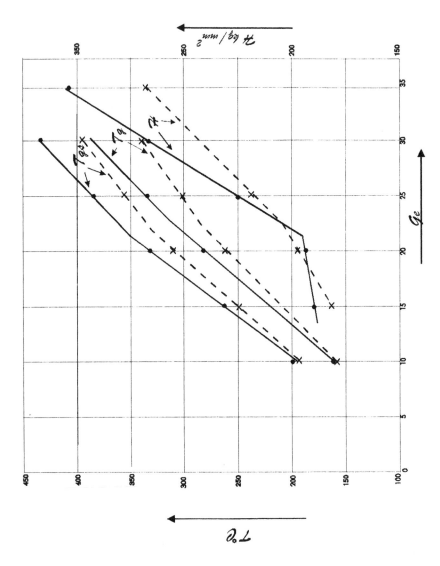

FIGURE 27. Property changes of glasses in the system (As_2S_3)–Ge–Se along the cut As = 20 at.% in comparison with the system As–Ge–Se. (—•—) As_2S_3–Ge–Se; (—×—) As–Ge–Se.

were the first to demonstrate that the sequential addition of iod considerably reduces T_{ds}.[114] Lin and Ho studied the composition influence upon chemical resistance of such glasses and established that iodine assists in its decrease.[115] The glasses, however, remain tolerant to moisture and acid solution actions, but, like other ChG, they get dissolved in alkalis. The authors attribute the influence of iodine additions on chemical tolerance to the structural changes. They suppose that, penetrating into glass, iodine infringes continuity of network by the sceme:

$$
\begin{array}{ccccccc}
-S-As-S- & & -S-As-S- & & -S-As-S- \\
| & & | & & | \\
S & \xrightarrow{\;I_2\;} & I & + & S \\
| & & & & | \\
-S-As-S- & & & & I
\end{array}
$$

(Str. 11)

Hopkins and others also think that in the system As–S–I there occurs the breakage of the bonds As–S and the formation of the bonds As–I.[116] However, sulfur atoms therewith become paired yielding the bonds –S–S, i.e., a laminated structure As_2S_3 sequentially gives way to a chain one. A bit later Borisova and Doynikov investigated the system As–Se–I and demonstrated that the insertion of iodine reduces microhardness and chemical tolerance of arsenic-selenium glasses.[117] T_{ds} of vitreous alloys $AsSe_xI_y$ varies between 150 and 40°C. The authors think that iodine disperses the structural network of glass.

Pearson and others investigated the system As–Te–I.[118] According to their data these glasses are hard at a room temperature and they get soft at 100 to 150°C, therewith some of them slightly crystallize. Recall that in the binary system As–Te there is no glass-forming under the ordinary overcooling conditions. Eaton investigated the properties of the glasses in the system As–Te–I in the transition region "glass-crystal" and discovered the changes in resistance from 10^3 to 10^7 Ω.cm.[119] Such a high value of electric conductivity makes us suppose that the author dealt with glass-crystalline alloys.

ChG containing bromine had been studied by Fischer and Mason,[120] and also by Pearson.[118]

A series of Deeg's works is devoted to the investigation of the system As–S–Cl.[121] The author describes in them the properties of glasses, methods of compositional analysis, and structural peculiarities.

Vitreous systems Ge–Se–I and Si–Se–I were described by Dembovsky.[122] Their low chemical stability was observed then.

In the 1970s there appeared the works by Chernov and Dembovsky,[123,124] — investigators of the iodine-containing systems, where they had come to the conclusion about the structure of these systems, and also a large work by Kirilenko and Dembovsky on the investigation of the glass-forming in the systems A^{IV}-B^{VI}-C^{VII}.[125]

Analyzing the presented literary data, one can see that the authors have studied only elementary ternary systems in laboratory conditions, working with a small amount of substances. This is little evidence for producing homogeneous materials satisfactory for application in optic devices. What's more, Pearson notes[126] that halogen-containing glasses can be considered only as laboratory curious, but not commercial material. In fact, some of these glasses are of a practical interest.

1.4.5.1 The Synthesis of Glasses with Halogens. The Method for Measuring the Softening Point

Halogen insertion demands certain changes in the synthesis process due to their enhanced volatility. The technique of founding in open vessels, applied, for example, by Pearsen and others[114,118] as well as Deeg,[121] inevitably results in selective evaporation of the components, whereas the glass compositions change. Besides, the poisonous vapors of arsenic compounds with halogens escape in the atmosphere therewith.

That's why in the 1970s the vacuum method of synthesis came into use. Thus, Johnson synthesized the glasses Ge–Se–I in the evacuated and unsoldered quartz ampules in the electric furnace.[127] Quin-Rod, investigating glasses in the system As–Te–I, also made them of elementary substances in quartz ampules. The operation of charge preparation can be performed in the atmosphere of pure dry argon.[128] The vacuum technique was also used by Khyiminetz and others,[129] while examining the glasses with bromine, inserted as the compound $SbBr_3$, which certainly limited the range of compositions.[129] However, the vacuum technique can also not be used unaltered for the synthesis of easily fusible and highly volatile materials, because the digesting vessels can then explode. That's why our initial aim was to work out the procedure satisfactory for the synthesis of ChG with halogens in a wide range of compositions.

Melnikov synthesized and investigated halogen-containing glasses. The alterations he introduced into the accepted synthesis technique include, primarily, the distinctive expedients of halogen insertions into the charge. This chiefly concerns liquid bromine and gaseous chlorine. For them special installations were made. For bromine insertion its calculated amount is accumulated in a glass buret with markings with the help of a rubber bulb with a regulating screw which is connected with a buret through a rubber pipe. A buret is put into the previously weighed digesting ampul and the necessary amount is introduced into it which is determined by the weight difference. The ampul with bromine is placed into liquid nitrogen and after slow cooling — in order not to let the reaction begin — other previously weighed elements are added. After that the evacuation is performed.

As chlorine reacts with halogens at room temperature, it is inserted directly into the charge. Among the multiple ways to recover elementary chlorine, the reaction of potassium permanganate interaction with concentrated hydrochloric acid, which goes at room temperature, appeared to fully conform to our aims. The installation consists of a retort with $KMnO_4$, plugged with a stopper through which the dropping fuel and the exhaust glass pipe go. Escaping chlorine comes into the digesting ampul where it immediately enters into the reaction with the previously prepared charge. Different amounts of chlorine may be inserted through an ampul's weighing at regular intervals. Further on an ampul is cooled in liquid nitrogen and the evacuation is carried out. Iodine is inserted into the charge without any special precautions, but to protect it from volatilization at evacuation, iodine is first filled into the digesting vessel and evacuation time is limited. Under manufacturing conditions, iodine losses from the vessel with a charge weighing 1 kg at its content equal to 400 g and 1-h evacuation is 1 g, i.e., 0.1% to the charge weight. Such an insufficient change of the composition doesn't essentially influence the properties of glass, which is confirmed by a good reproducibility of properties of glasses of commercial meltings.

Ampules with bromine and chlorine containing charge are pumped out with permanent cooling by liquid nitrogen and transfer halogens to the solid state and also substantially eliminate the losses. The advanced technique of the charge preparation is comparatively easy, which allows one to obtain halogen-containing glasses of high quality and any composition. The time-temperature synthesis regime is worked out for each glass type, respectively, depending on the halogen nature, with regard to the fact that iodine and, especially, bromine and chlorine not only reduce the viscosity of glasses to a large extent, but also lead to rise of the vapor's pressure over fusion, which scales up the possibility of melting vessels' explosion in the process of melting. That's why the accepted maximum synthesis temperature is equal to 500 to 700°C, depending on the content of halogen. In the systems containing the elements of the IV group it may be increased up to 800 to 900°C. While the temperature is rising, holding at 200°C is 1 h and at the clarification temperature — 2 to 4 h. Mixing occurs for 30 min every hour. After the furnace is disconnected spontaneous cooling proceeds at 100 to 200°C per hour up to 300 to 500°C (depending on the infusibility of glasses) and after that the ampul with the glass is quenched to room temperature.

Halide glasses possess some distinctive properties which demand peculiar methods for their measuring. First of all is the low melting point. It will be recalled that some compositions with maximum content of halogen keep in plastic or liquid state at room temperature. It leads to certain experimental problems, especially when at measuring T_{ds}. Usually an applied dilatometric method is unsuitable for the diversity of easily fusible glasses with T_{ds} below 50°C. To measure T_{ds} of easily fusible and crystallizing glasses we have worked out the method which represents the modification of the pressing-in method applied at vicousity measurements. For measurements we did not use the standard viscosimeter, but the device we had specially constructed (Figure 28a). The lifting screw (1) for indicator and indentor is fixed on the mount. Inside the electric furnace with thermal insulation (2) a brass cylinder is placed for the temperature leveling in the furnace combustion space; its size Ø 32 × 70 mm (3). A flat-parallel specimen (4) is set up on a quartz stand (5) with 0.5- to 1-mm clearance with cylinder walls and bears up against the quartz pipe (6) and the mount. An indentor (7) connected with an indicator (8) is installed on the specimen. It should be noted here that the system of quartz details bears up only against the mount, and at the heating up, due to the low quartz expansion, it demonstrates the specimen expansion without any additional errors. The temperature is determined by the thermometer or thermocouple (9), inserted into the heat-insulating cover of the cylinder with openings for thermometer and indentor. Prior to measuring under the action of an indicator spring, a fixed load is set of approximately 100 g (on a small scale — 0.2 μm). The typical impression depth change-temperature curve is presented in Figure 28b. At the intersection of the lines ab and cd we can find T_{ds}. The section cd is rectilinear in the impression depth range of 30 to 100 μm. For the glasses with T_{ds} below room temperature the installation is offered which contains a double-walled quartz tumbler for the prior cooling of the substance being measured by liquid nitrogen. The quartz tumbler is placed into the heat-insulating case with a foam cover. Liquid nitrogen is introduced between the walls of the glass. Due to good insulation the necessary speed of the temperature rise in the working chamber equal to 1 to 2°C/min is attained spontaneously. Besides, the speed can be regulated by adding liquid nitrogen.

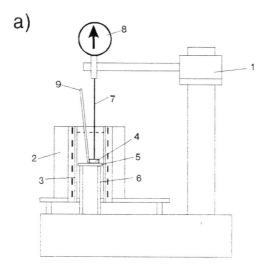

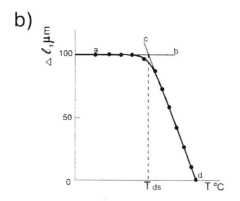

FIGURE 28.   (a) Installation for T_{ds} measuring by the impression method; (b) alteration of the depth of indentor's impression due to temperature.

For the measurements taken by the above-described methods, the high accuracy of the mechanical treatment of the specimen surfaces isn't necessary, because the globule on one end of an indentor is self-adjusting and that makes the device insensitive to curvings. Both polished and ground or prepared by hot forming specimens can be used shaped as flat-parallel plates $2 \times 5 \times 5$ mm. The glasses which are liquid at room temperature (for example, those of the system As–S–Br) are dropped on the bottom plate and cooled in the measuring chamber until hardening and after that the measurements are carried out in a usual way. Small sizes of specimens which demand 0.1 to 0.5 g of substance and the possibility not to put them through the repeated thermal treatment, necessary at molling and annealing, permitting measurement of readily crystallizing glasses. Low impression depth (100 μm) enables one to determine T_{ds} for thin films, covers, etc. Since the measuring period for temperatures up to 200°C isn't over 1 h, express investigations can be carried out at this device. In Table 3 T_{ds} of the glasses produced by the method described is given in comparison to measurements on a standard dilatometer. One can see from the table that the results essentially agree. We must also keep in mind that the change of the weight of the charge components, especially of such highly volatile ones as bromine and iodine, gives much greater scatter in the data, not providing the accuracy of the device (±1°C).

TABLE 3
Comparison of Softening Point Obtained by the Methods of Pressing in and Dilatometry

Composition in at.%	Softening point (°C)	
	Pressing in	Dilatometry
$As_{40}S_{60}$	212, 212, 212	210[a]
$As_{19}S_{34}Br_{47}$	−63, −63, −63	−60[118]
$As_{20}S_{60}I_{20}$	15	—
$As_{30}Ge_{30}Se_{40}$	385	387[a]
$As_{20}Ge_{20}Se_{60}$	295	294[a]
$As_{10}Ge_{10}Se_{80}$	135	135[a]
$As_5Ge_5Se_{90}$	91	90[a]

[a] Our measurements.

Since with halogen insertion the electron absorption edge shifts to a short-wave region, the control over the monophase character of glasses is not a problem. The presence of crystalline or amorphous insertions is clearly viewed under the IR microscope. Only for compositions with low content of halogens the monophase character is determined according to the type of transmission spectrum.

1.4.5.2 Properties of Halogen-Containing Glasses

We have studied the glass-forming and properties of halogen-containing glasses in the following systems:

- With iodine: As–S–I; Sb–As–Ge–S–I; As–Se–I; As–Ge–Se–I; Sb–As–Ge–Se–I; As–Te–I
- With bromine: As–S–Br; Sb–As–Ge–S–Br; As–Se–Br; As–Ge–Se–Br; Sb–As–Ge–Se–Br
- With chlorine: As–S–Cl; As–Ge–S–Cl; Sb–As–Ge–S–Cl

1.4.5.2.a   Glasses of the Systems As–S–I and Sb–As–Ge–S–I

The glass-forming region in the system As–S–I essentially coincides with those discovered by Pearson.[114] All the glasses are transparent in the visible. The glasses containing a large amount of arsenic are colored dark red, which turns bright red as the iodine content increases. Therewith T_{ds} comes down so that some glasses are plastic at room temperature. Actually they are the melts, which at deeper cooling gain the properties of hard glasses. The crystallization interval therewith can include room temperature, and such glasses crystallize upon storage. The crystallization period lasts from 1 to 2 d to several months, depending on the crystallization speed and temperature. At the second melting the crystallized samples easily transfer into the vitreous state. On addition of iodine at the expense of sulfur, the density of glasses naturally enhances, but atomic concentration considerably decreases and that is evident of structural change with the breakage of continuity of the vitreous network and its disintegration owing to the large size of an iodine atom.

The complex system As–Sb–Ge–S–I represents, in fact, one of the cuts of the system As–Ge–S, where 10 at.% As is replaced with 10 at.% Sb, and also 10 at.% I is inserted at the expense of sulphur. In the subsystem Ge–S–I the glasses are formed in the range from 10 to 20 at.% Ge, and they are light red. In the glasses of composition including Sb, and also Sb + As, with the enhancement of Ge content the color changes from light red to black and T_{ds} considerably increases (from 158 to 333°C).

1.4.5.2.b   Glasses of the Systems As–Se–I and
*   Sb–As–Ge–Se–I*

In the system As–Se–I the glass-forming region is substantial and limited by the compositions ~60 at.% As and ~45 at.% I. The size of the glass-forming region agrees in essence with the data by Borisova,[117] but, as opposed to the authors, we didn't observe any markers of stratification. A series of glasses is really crystallized — completely or partially — in storage, but, as well as in sulfide glasses, the crystallization is determined by coincidence of its range with room temperature. On the addition of iodine T_{ds} decreased, from 160 to 20°C. We obtain complex glasses based on the system As–Ge–Se through the replacement of 10 at.% As with antimony and 10 at.% I insertion in place of selenium. On the addition of 10 at.% I to the principal system dark-red glasses are produced; after arsenic replacement with antimony the color doesn't essentially change.

The glass-forming region and softening temperature surface in the system As–Ge–Se–I are demonstrated in Figure 29. The insertion of iodine lowers T_{ds} of the glasses of the principal system, but they are still infusible enough; T_{ds} of glasses with antimony additions is slightly lower than of pure arsenic ones. In Figure 30 transmission spectra of selenium glasses with iodine are shown. The values of the properties of all glasses of these systems are given in Table 3 of the Appendix.

1.4.5.2.c   Glasses in the System As–Te–I

On iodine insertion into the system As–Te — initially a nonglass-forming one — vitreous alloys were obtained outwardly with conchoidal fracture, with intense metal

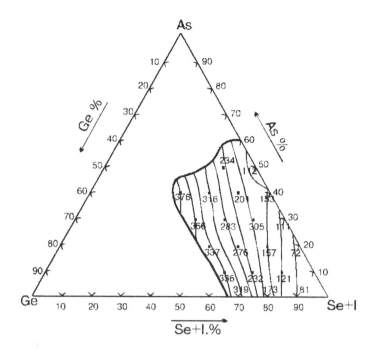

FIGURE 29. Glass-forming region and T_{ds} surface in the system As–Ge–Se–I.

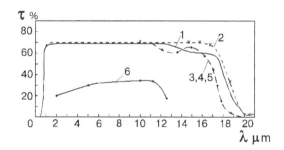

FIGURE 30. Transmission of iodine-containing glasses: (1) $As_{10}Se_{90}$; (2) $As_{20}Se_{50}I_{30}$; (3) $As_{20}Ge_5Se_{70}I_5$; (4) $Sb_{10}As_{10}Ge_5Se_{70}I_{15}$; (5) $Ge_{10}Se_{70}I_{20}$; (6) $As_{45}Te_{40}I_{15}$.

glance. At 10 g suspensions with quenching from 500 to 700°C such alloys are formed within 43 to 52 at.% As and from 5 to 35 at.% I. They are opaque even under the IR microscope and their monophase character can be valued only by the transmission spectra. Since the maximum transmission of the samples 2 mm thick was equal to 30 to 40% and the spectrum curve was flat (Figure 30), it came absolutely obvious that the samples were partially crystallized. At storage the alloys become coated with the crystalline film, i.e., they are chemically unstable particularly to damp atmosphere. Hence in the system with halogen pure tellurium glasses are of no practical interest.

1.4.5.2.d   Glasses of the Systems As–S(Se)–Br(Cl) and
Sb–As–Ge–S(Se)–Br(Cl)

Insertion of bromine dilutes glasses to a large extent; part of them are plastic at room temperature, and multibromine (heavy-bromine) glasses remain liquid. Thus, T_{ds} of the glass $As_{19}S_{34}Br_{47}$ is equal to $-63°C$. Especially engaging our attention is the low chemical activity of the glasses with a large bromine content: they can be stored in liquid state in an open vessel for several years without getting oxidized and crystallized. A part of fusible glasses with the crystallization interval coinciding with room temperature is crystallized at storage. The glasses with bromine are colored light green or yellow. Complex glasses here, like in all other cases, are obtained through 10 at.% As replacement with antimony and 10 at.% Br insertion at the expense of sulfur. Glass-forming region is bounded by the compositions 25 to 50 at.% As + Sb and up to 30 at.% Ge. There are glasses in the subsystems Sb–Ge–S–Br and As–Ge–S–Br. Glasses in the ternary system Ge–S–Br were never obtained.

As the content of arsenic increases, staining changes from red to dark red, and as the content of germanium increases, from dark red to black.

It is much more difficult to introduce bromine into the system As–Se than into the sulfur one: stable glasses are formed only close to the line As–Se. All glasses are black; a part of them becomes coated with a crystalline film at storage. A few glasses are also formed at bromine insertion into the system As–Ge–Se, and at arsenic replacement with 10 at.% Sb glasses are formed only in the absence of germanium. It is possibly connected with the fact that in our method the composition is being held more accurately due to a rather low volatilization of components. We obtained in vitreous state the compositions containing from 10 to 45 at.% As and up to 25 at.% Cl.

On germanium insertion into the known system As–S–Cl and on 10 at.% As replacement with antimony and 10 at.% S replacement with chlorine complex glasses are formed in the system Sb–As–Ge–S–Cl and also in the subsystems Sb–S–Cl, Ge–S–Cl, and As–Ge–S–Cl. All glasses are light yellow. The compositions of the glasses containing bromine and chlorine and T_{ds} are given in Table 4 of the Appendix.

1.5. PROPERTIES OF THE GLASSES AS A FUNCTION OF AVERAGE NUMBER OF COVALENT BONDS PER ATOM (THE COVALENT ATOMIC CORRELATION)[18–20]

To compare the values of microhardness in the systems Ge–Se and As–Se, one can see that microhardness of germanium glasses is considerably higher than that of adequate atomic composition glasses with arsenic. This seems surprising, because atoms of arsenic and germanium have almost equal covalent radii and very similar values of electronegativity and bond energy with selenium. As long ago as 1959 we supposed that covalent atomic correlation influenced the properties of glasses and put forward the formula for its calculation.

$$K_{gl} = \sum_{1}^{i} \frac{K_i A_i}{200}$$

where K_{gl} is average covalent atomic correlation, K_i is number of covalent bonds formed by an atom of a given element, and A_i is content of a given element in glass in at.%. K_i had been calculated on the assumption that arsenic and its analogues in the subgroup form three bonds; elements of the IV group — four bonds; and chalcogens perform bridge connectors, i.e., they are doubly connected. The calculations which had been made for binary systems demonstrated that glasses with equal values of covalent correlation have equal values of microhardness (Figure 31).

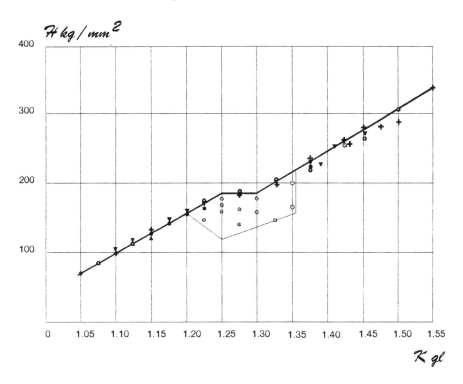

FIGURE 31. Microhardness change in the glasses of different systems depending on the covalent correlation of atoms.

Later on we passed this mechanism as applied to the thermal properties of glasses of elementary and complex systems and it was verified by Nemilov on the example of viscosity of glasses in the system Ge–Se.[134] Changes in thermal properties of the studied glasses are shown in Table 1 of the Appendix and in Figure 32. The deviations in property values for glasses with equal covalent correlation in the region with chalcogen excess are not higher, on the average, than 5% of the measured value. Taking into account that the measurements had been carried out by various methods, various measurers, and on the specimens of various meltings, such agreement may

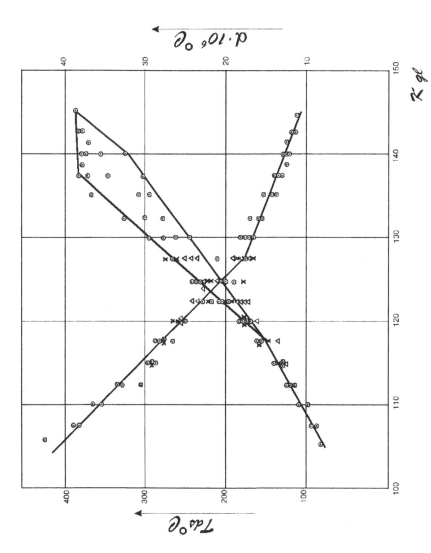

FIGURE 32. Measuring of T_{ds} and α of the glasses from different systems depending on the covalent correlation of atoms.

be considered to be quite satisfactory. At the same time one can see from the pictures that in the region with a lack of chalcogene, where the bonds As–As and Ge–As appear, the dependence on the number of bonds remains only for α and microhardness. Values of T_{ds} and T_g depart noticeably from this mechanism.

Analyzing the experimental material, we have come to the conclusion that the reason has to do with the energy parameters of resulting bonds, because the properties of glasses, determined by the bonds' breakage or changeover, must depend not only on their number, but also on energetically diverse structures. True, in the region with the lack of chalcogene in spatial glassy network with the bonds' energies 52 to 54 kcal/mol (Ge–Se and As–Se), the structures with the bonds' energies 46 to 47 kcal/mol (As–As and Ge–As) appear and increase in number. To check the correctness of this reasoning for all investigated glasses, the sum of partial energy values was calculated, introduced by each bond type, which can be called an average energy correlation of atoms. The values of homobonds' energy, taken from the literature,[135] and heterobonds' energy calculated, considering electronegativities, with Poling's formula,[68] are presented in Table 4. In Figure 33 one can see that within the accuracy of the experiment and calculation the values of T_{ds} of all glasses fit on one line with fracture at the transfer into the region of structures with the lack of chalcogene. Microhardness depends mainly on the concentration of bonds and that determines departures in the region where the bonds As–As are predominant; α is measured at temperatures which allow thermal fluctuations to occur without bonds' changeover. These properties don't respond to the changes of the energy parameters of the system. These data were published in the magazine "Optic-Mechanical Industry" in 1961 and in the collection "The Structure of Glass" in 1969.

TABLE 4
The Energy of Unitary Bond
Breakage (kcal/mol)[135]

	S	Se	As	Sb	Ge	Sn
S	65	—	—	—	—	—
Se	57	49	—	—	—	—
As	61	52	46	—	—	—
Sb	62	51	44	42	—	—
Ge	68	54	47	44	46	—
Sn	60	51	42	39	41	36

In 1979 Philips paid attention to the importance of the concept of covalent correlation of atoms in glass. However, from our point of view, it wasn't valid enough when he connected this concept with the glass-forming ability of substances.[17] His topologic glass-forming criterion is calculated by the formula

$$N_g = m/2 + m(m-1)/2 = m^2/2$$

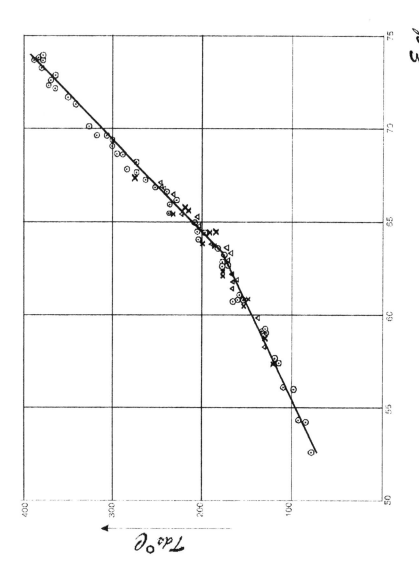

FIGURE 33. T_{ds} change depending on the energy correlation of atoms.

where m is an average coordinating number. Since in the majority of ChG the coordination number coincides with the number of valene bonds, the value N_g accounts for covalent correlation of atoms. While calculating m the author used coordinating numbers equal for Ge-4, As-3, and Se-2, which is correct for the most part, and this enables one to separate nonglass-forming chalcogenides of heavy elements from the glass-forming ones. However, the maximum glass-forming tendency, after Philips, answers the value $N_g = 2.45$, that isn't right, in fact. It is known that even in the binary systems As–Se and Ge–Se the most stable glasses are formed in the region of noneutectic compositions, N_g of which is much lower. At the same time the dependence of glass properties on the average number of covalent bonds, accounted for one atom, is beyond question. Covalent correlation may be replaced with the average coordination number, certainly only in the systems where these values agree.

In the last few years a series of works has been published in which an average coordination number, properties, and structure of ChG are held together. The work investigating the systems As–Se, Ge–Se, As–S, Ge–S, and As–Ge–S, carried out by different authors, was published in 1989 by Tanaka.[136] The author established that such properties as atomic volume, density, elastic modulus, and reversible photoeffect depend on the average coordinating number.

In the 1990s publications appeared in which the works by Philips and Tanaka were cited. For example, Giridhar and Mahadevan established T_g in the system Ge–In–Se (In = 5 and 8 at.%) dependent from the average coordinating number.[137] Tziulyanu and Gumenyuk,[138] while investigating the system As–S–Ge in the region with sulfur excess and calculating the average coordinating number in the same way as in previous works, called this value "average number of covalent bonds per atom". They had established the dependence of optic width of the forbidden zone and the molar volume of r^-.

Noted separately should be the works by Sreeram and others, published in 1991, because the authors had been studying the complex quinary system Ge–Sb–As–Se–Te and also the subsystems Ge–Sb–Se and Ge–Sb–Se–Te, where sufficiently stable glasses can be formed.[139] Data for Vickers microhardness and for T_g were obtained. The results are reserved depending on the average coordinating number. As could be expected in the region with excess of chalcogene, the values of glasses' properties with one and the same r agree, and in the region with the lack of chalcogene the disagreement is observed. The fact that in glasses of different systems with the same coordinate number, the values of the properties are the same, can hardly be explained. However, it comes clear if originating from the middle covalent or energetic correlation of atoms.

Unfortunately, the property dependences from the covalene correlation of atoms isn't universal (and hence, in particular, it can't serve as a criterion of glass-forming). It is observed only in the systems with pure covalent bonds. When ionogene components appear — in ChG they are thallium and halogenes — pair-electron bonds break, forming dipoles and quadrupoles, as, for instance:

$$
\begin{array}{c}
- \text{Se} \\
\backslash \\
\text{As} - \text{Se}^-\text{Tl}^+ \qquad \text{Se} - \\
/ \qquad\qquad\qquad / \\
- \text{Se} \qquad \text{Tl}^+\text{Se}^- - \text{As} \\
\backslash \\
\text{Se} -
\end{array}
$$

(Str. 12)

Such bonds can't be considered completely broken due to quadrupole interaction, the energy of which isn't determined. That is why we can't apply this concept to oxide glasses, the majority of which contain ionogene elements.

However, the fact that the experimentally found values of ChG properties, so different in compositions and components' nature, coincide quantitatively if based on covalent correlation of atoms, provides impressive evidence that the concept of glasses as covalent correlated polymers with enriched valene bonds and chain, laminated or ternary structure, is right.

2 Optical Properties of Chalcogenide Glasses[21-23]

2.1 SPECTRAL TRANSMISSION, ABSORPTION BANDS

In the present work, to achieve the principal aim of creating optical medium for IR-spectrum range, the investigation of spectral characteristics was carried out in the interval of the maximum transparency of these glasses, i.e., at wavelengths from 1 to 25 μm.

The absence of crystalline or any other imperfections in the glass volume doesn't prevent the appearance of extrinsic absorption bands, due to the presence in glass of the elements which perform bonds with greater oscillation frequency than is inherent in the principal network. At the same time the manufacturing of oxygen-free glasses, transparent beyond 14 μm, meets problems stemming from the intrinsic absorption bands. That's why the analysis of the band origin is of paramount importance.

The distinction between intrinsic absorption bands, deriving from the oscillation of glass bonds, and extrinsic bands can be performed on the grounds that the intrinsic absorption band must be observed in the spectra of glasses of the given compositions, independently from the purity of the initial materials and changes in the synthesis regime. Its intensity in a glass of a given composition must remain practically constant. If any of the bands is absent in the spectrum of at least one specimen, it should be considered extrinsic.

It is advisable to begin the analysis of absorption spectra with the examination of vitreous selenium and glasses of the elementary systems based on it. Pure vitreous selenium is transparent up to 20 μm. Gebbie and Cannon observed in two absorption bands in its spectrum: one weak at 13.5 μm and one more intensive at 20.6 μm, which they considered to be intrinsic.[140] This was confirmed in the works by Ballard,[141] Vasko,[142] and also in our research.

Along with intrinsic bands in selenium spectra, one can often meet a great amount of extrinsic ones, caused by the presence of moisture and oxygen.

Spectral characteristics of different batches of selenium, produced by manufacturers, are, as a rule, very different. The purity of these batches for controlled metal admixtures is identical and sufficiently high. Nevertheless, in all specimens bands of different intensity at $\lambda = 7.6$, 8.8 to 9, and 13.5 μm are observed. In addition, they all are of lowered general transmission due to foreign inclusions clearly viewed under the microscope. In Figure 34 one can see how the spectral purity of selenium influences the characteristics of complex glasses, synthesized on its basis.

Systematic investigations of spectra of the glasses As–Se, Ge–Se, and As–Ge–Se were first begun in our laboratory and we had been carrying them out since 1957. The specimens for optic property investigations were put to the test for purity and homogeneity. The thickness of specimens was from 1 to 10 mm. Transmission in the region of continuous electron spectrum edge was measured with spectrophotometers SPh-4, Unicam, and Hitachi ESP-3; in the range between 1.1 and 25 μm — with the single-ray spectrophotometer IRS12, and the double-ray Hilger-Watts H-800 and UR-10. In 1957 to 1961 it had been demonstrated that the glasses of all three systems are high transparent in the interval of wavelengths 1 to 15 μm.

In the glasses of binary systems and of the system As–Ge–Se with excess of selenium there was, as a rule, a band of various intensity observed in the wavelength interval 12.5 to 13 μm with the center 12.8 μm, removed at the correct synthesis regime. In the glasses As–Ge–Se with the lack of selenium the band at 12.8 μm was always observed and the changes in the synthesis regime produced only slight variations of intensity. However, Yefimov managed to prove the extrinsic origin of this band in glasses with lack of selenium, having removed it from the glass $As_{15}Ge_{30}Se_{50}$ by additional purification of arsenic, and from the glass $As_{15}Ge_{10}Sn_{10}Se_{65}$ by adding activated carbon. We should note that such a technique of the charge release from oxygen can't be recommended for commercial application because the band at 4.5 μm appears here (probably, CO_2 admixture) and black insertions of carbon, which lower the total transmission, remain in glass. It became obvious that the band 12.8 is caused by the presence of oxygen which reacted with As and Ge during synthesis, i.e., by the appearance of bonds As–O and Ge–O. The coincidence of absorption bands in the spectra of the glasses As–Se, Ge–Se, and As–Ge–Se is explained by the proximity of arsenic and germanium atomic weights. In the spectra of pure oxides As_2O_3 and GeO_2, the position of the first intensive bands (respectively, 12.5 μm and 11.5 μm) is close to the position of the examined band. Along with the band with center 12.8 μm, incidentally appearing bands were also registered at 7.9 to 8.0 μm in glasses containing germanium and at 9.5 to 9.6 μm in the glass As–Se, and also the band 6.3 μm which is always followed by the weak band 3.3 μm. Besides, it was established that at the most accurate purification of the charge materials the presence of oxygen in the melting vessel is sufficient for the appearance of a strong band 8.0 μm, with the considerable transparency decrease in the region 8 to 12 μm. Furthermore, in the spectra of all the glasses As–Se, Ge–Se, and As–Ge–Se, there exists a wide, intensive, absorption band in the long-wave part of the spectrum, which causes the transmission to decrease to ≈0 in the specimens 2 to 3 mm thick. This band for As–Se glasses is located in the wavelength interval 20 to 21 μm and for Ge–Se and As–Ge–Se, glasses 18 to 19.5 μm.

Comparison of the spectra As_2O_3 and GeO_2 with the spectra of investigated glasses enables one to explain the origin of these bands. Really, the bands at 9.5 to 9.6 and 20.6 μm take place in As_2O_3 spectrum, and in the GeO_2 spectrum the band at 18.0 μm is observed. Nevertheless, the intensity of long-wave bands remains constant in all the specimens and it doesn't depend on the presence or absence of

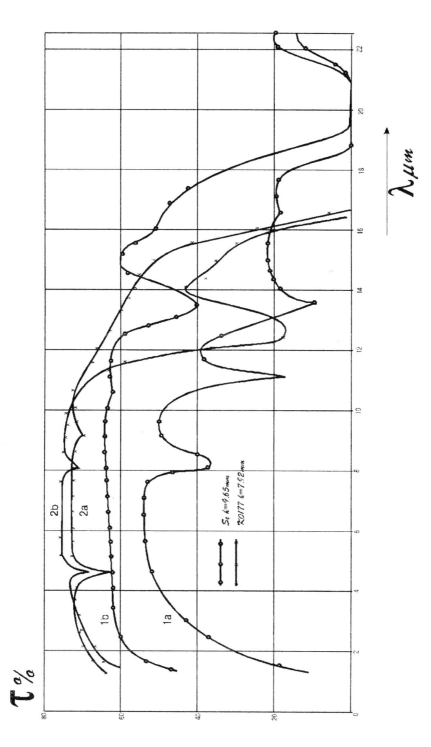

FIGURE 34. Spectral transmission of selenium, unpurified (1a) and purified in the laboratory (1b) and of the glasses based on it (2a and 2b).

an oxide band at 12.5 to 13.0 μm. From this it follows that these bands are intrinsic absorption ones of the glass network.

To establish the nature of extrinsic absorption bands, two glasses — Ge_5Se_{95} and $As_{40}Se_{60}$ — were synthesized in 30 g batches in the following conditions:

1. With water: prior to the evacuation the charge in an ampul was moistened by 0.5 sm^3 of water.
2. With atmosphere oxygen: an ampul with the unevacuated charge.
3. With oxide insertions: in germanium glasses 0.004 mass% of germanium oxide was added; in arsenic ones, 1.0 mass% of arsenic oxide.

The results appeared to be somewhat unexpected:

1. Instead of a water band in glasses melted from the wet charge, only a slight decrease of total transmission was observed. Such a rare dissimilarity of ChG from oxide ones is explained by their inertness to water.
2. Atmosphere oxygen produced the following absorption bands: in glass Ge_5Se_{95} — 2.8, 4.2, 6.3, 7.3, 11, 12.8, 15 μm; in glass $As_{40}Se_{60}$ — 2.8, 8.5, 12.8, 15.3 μm.
3. Oxygen, added with the charge, produced in germanium glass bands at approximately the same wavelengths, but of lower intensity. Oxide was inserted into arsenic glass in large amounts and caused a drastic transparency decrease throughout the spectrum, probably due to a partial crystallization.

Based on our investigations we came to the conclusion that all the absorption bands noted were caused by extrinsic oxygen; therefore, for rather deep bands to appear, we needed only atmosphere oxygen to enter the glass melt. Its insertion with the charge might lead to abrupt transmission decrease up to the absolute opacity.

Further investigation of the spectra has demonstrated that germanium is responsible to a large extent for the absorption band 12.8 μm. Even in cases when arsenic is an oxygen donor, disproportionation of bonds occurs in the process of synthesis, because oxygen (as well as sulfur and selenium) enters into the reaction first and foremost with germanium. According to the positions occupied by these elements in the periodic system, absorption bands of germanium-free arsenic glasses are shifted toward greater wavelengths.

The proposals about the origin of bands in the region 20 to 21 μm that we have come up with are confirmed by other authors, for instance, by Vashko[143] and Edmond and Redfearn.[144] Therewith the diversity of their opinions is explained by the fact that each of them takes into account only one of the causes, producing the absorption in the indicated part of the spectrum. Besides, in glasses with the greatest content of selenium its own absorption band 20.6 μm, described by Gebbie and Cannon.[140] must be superimposed here. The second intrinsic absorption band 13.5 μm also clearly manifests itself in these glasses.

In glasses of the system As–Ge Edmond and Redfearn noted the existence of one more weak band with diffused edges, which is located around 14.3 μm.[144] By analogy with the spectrum As$_2$S$_3$ where one can see two intrinsic bands, the strong one 14.2 µm and the weak one 10.0 µm, the authors considered the band 14.3 µm to be similar to the band 10.0 µm, i.e., to be the second intrinsic band of arsenic selenide. Such speculation is quite credible because they used rather pure specimens and the intensity of several bands of an obviously extrinsic type, which they had observed, was much lower than the intensity of the band 14.3 µm. Probably, the presence of this exact band can explain the decrease of glasses transmission after 12 to 13 µm, determined by us. In other works this band couldn't be discovered because of the camouflaging influence of adjacent intensive extrinsic bands at 12.5 to 13.0 and 15.7 to 16.0 µm.

True enough, in a 1982 article Savage pointed out only one intrinsic band in the glasses As–Se, at 21.4 µm, and three extrinsic bands deriving from the oscillations of the bonds As–O, at 9.5, 12.8, and 16.0 µm.[105] The position of absorption bands in the spectra of the glasses Ge–Se and As–Ge–Se, established by Savage and Nielsen in their early works[131,145] and also in the articles by Savage and Savage and Webber,[105,146] dated 1980 to 1982, is fully confirmed by our data.

The spectra of the initial glass systems without extrinsic absorption bands are presented in Figure 35. It should be noted that the transmission at wavelengths more than 12 µm can change with the composition in the range of several percent.

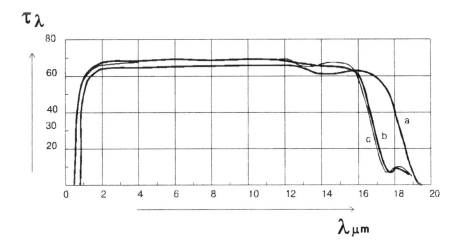

FIGURE 35. Transmission spectra of glasses from the systems As–Se (a), Ge–Se (b), As–Ge–Se (c) without extrinsic absorption bands (thickness 2 mm).

Among the intrinsic bands stemming from the principal components of glass in the systems containing arsenic is the band in the region 20 to 21 µm, which represents superposition of at least two intrinsic absorption bands from Se–Se bonds (in the glasses with the excess of selenium) and from As–Se bonds.

The intrinsic absorption band, caused by oscillations of the bonds Ge–Se, exists at $\lambda = 19.3$ µm. In the system As–Ge–Se spectra are of the intermediate character — between germanium and arsenic ones. The spectra in the regions with selenium excess do not differ a lot from those in the region with its lack. However, the extrinsic band 12.8 μm is much more difficult to remove from the last. One can think that oxygen here plays a role of a bridge between two atoms of arsenic and germanium, thereby connecting up more tightly than in selenium chains.

In the spectra of complex glasses produced with partial replacement of the components of the principal system As–Ge–Se by their more heavy analogues in the subgroup, both intrinsic and extrinsic absorption bonds must shift to the long-wave part of the spectrum. True, in the spectra of glasses of the As–Ge–Sn–Se system the band 12.8 μm manifests itself much more rarely, and in the systems As–Sn–Se and As–Sb–Sn–Se it is practically never observed. Sufficiently pure glasses, synthesized without deviations from the technical process, are characterized by only one intensive band at 20 to 21 μm, caused by the bonds As–Se and Se–Se. Insignificant transmission decrease near 14 μm is apparently connected with the diffused band, observed by Edmond and Redfearn in arsenic selenide.[144] At addition of 0.01 mass% SnO in the spectra of the glasses As–Sn–Se and As–Sb–Sn–Se the absorption band at 14.5 to 15.0 μm appears — apparently due to oxygen correlated with tin. Extrinsic bands caused by oxygen were removed with the help of activated carbon, inserted into the ampul prior to synthesis in the amount of 0.01 to 0.02 g per 10 g of glass. It should be noted that even in the case of high intensity of the band at 14.8 to 15.0 μm the spectral curves of the glasses As–Sn–Se and As–Sb–Sn–Se in the region closer than 12 μm don't suffer any alterations as a rule.

In Figure 36 the spectra are shown of five 10 mm thick specimens of the glasses of the As–Sb–Sn–Se system of different meltings with successively increasing content of oxygen. One can see that the band 12.8 μm intensity is considerably lower than 14.8 to 15.0 μm. Hence in these glasses oxygen in the main binds with tin.

In the glasses of the As–Ge–Sn–Se system oxygen probably binds equally with tin and germanium, hence two absorption bands are observed here: 12.8 μm and 14.8 to 15.0 μm, i.e., partial germanium replacement with tin can't provide complete removal of the band 12.8 μm. The above-mentioned also concerns the glasses of the As–Ge–Pb–Se system.

In the spectra of glasses of the As–Ge–Pb–Se system the absorption band 12.8 μm is observed more rarely and its intensity is slightly lower, but it's more difficult to remove it from here than from the tin glasses. On the addition of lead no new bands are observed. Apparently, lead only slightly binds oxygen as it could be expected according to the comparative value of Gibbs free energy of its oxides' potential.

Reasoning from the foregoing let us present the list of absorption bands observed by Yefimov in glasses of the investigated systems.

Removing the above-noted extrinsic bands from glasses allows one to use them successfully in thermovision optic systems up to 17.5 to 20 μm.

In the 1980s in connection with ChG application in fiber optics, works appeared dedicated to the investigation of optic losses in fiber. It allowed the degrading of some extrinsic bands into the components to find new ones which don't manifest themselves at usual spectral measurements and to refine the position of the known bands. Thus, the wide band in the 2.7- to 2.9-μm region appeared due to the OH⁻ group and H_2O, which we found out in transmission spectra was degraded by

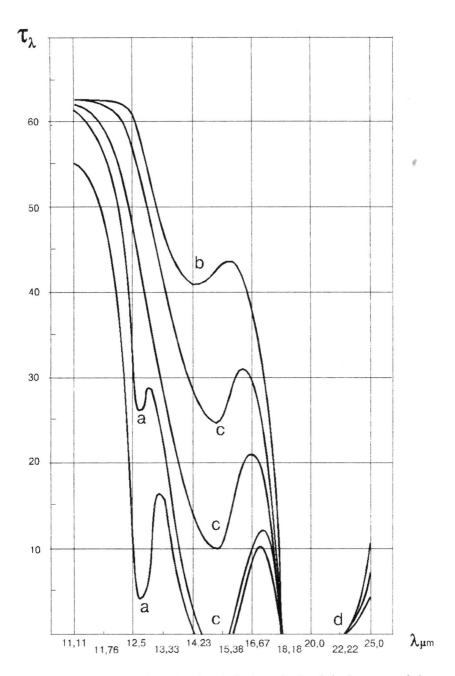

FIGURE 36. Ratio of the intensity of extrinsic absorption bands in the spectra of glasses of the system As–Sb–Sn–Se with different oxygen content (thickness 10 mm). (a) Extrinsic band from As–O bonds (12.7 to 12.9 μm); (b) intrinsic band from As–Se bonds (14.3 μm); (c) extrinsic band from Sn–O bonds (14.8 to 15.0 μm); (d) absorption from bonds of the glass network.

Extrinsic Bands

Position (μm)	Origin	Intensity
2.7–2.9 } 6.3	OH⁻	Weak
4.5–4.9	H–Se	Weak
7.9–8.0	Ge–O	Weak
9.5–9.6	As–O	Weak
12.5–13.0	Ge–O and As–O	Intense
14.8–15.0	Sn–O	Intense
20.0–21.0	As–O	Intense

Intrinsic Bands

Position (μm)	Origin	Intensity
13.5–13.6	Se–Se (in chains)	Weak
14.2–14.5	As–Se	Weak
17.8–18.0	Ge–Se	Intense
20.0–21.0	Se–Se (in chains) } As–Se	Intense

Kanamori into the contents with overtones at wavelengths 1.92 and 2.2 to 2.3 μm.[147,148] They also found the bands at 3.55 and 4.15 μm explained by the presence of the complexes

$$> As - Se - H \qquad\qquad (Str.\ 1)$$

and the bands deriving from vibrations of S–H bonds (4.01 to 4.1 μm with overtone 2.05 to 2.1 μm) and the bands which can be explained by the complexes

$$> As - S - H \qquad\qquad (Str.\ 2)$$

(2.55, 3.09 to 3.11, 3.64 to 3.69 μm) confirmed in the work by Borisevich and others.[149] In the work by Kanamori the overtone 2.3 μm was ascribed to the principal band 4.7, corresponding to H–Se oscillations.[147] When the amount of hydrogen is large in selenium, the bands 7.17 and 14.4 μm appear, revealed in the works by Skripachov and others.[150,151] In the work by Moynihan and others the band 15.9 μm is also ascribed to hydrogen combined with selenium.[153] According to the data by Savage and others,[146] the band at 4.9 μm corresponds oscillation frequency of the bond H–Te. The position of the second water band at 6.3 μm is confirmed in the works by Bychkova and others.[152] This is the extensive group of extrinsic absorption bands connected with hydrogen presence in glasses.

The number of bands, having their origin in vibrations of the bonds of principal components of oxygen glass and located in the 7.5 to 21.0 μm spectral region, is also quite large. They are the most detrimental to IR-optics, for they straddle almost entirely the third atmospheric window. The greatest number of bands is ascribed to the oscillations of As–O bonds in various forms — from As_4O_6 to the mixed –S–As–O–.

Besides 9.5 to 9.6, 12.7 to 13.0, and 20.0 to 21.0 µm, revealed by us, Moynihan established in As_2Se_3 the presence of the bands 7.5, 7.9, 8.9, and 10.4 µm,[153] and Lezal the bands 8.6, 9.1, 10.2, and 10.7 µm.[154] Webber and Savage also ascribe the 14.1 µm band to the oscillations of As–O bonds.[86] We ascribe the bands 7.9 to 8.0 µm to the oscillations of Ge–O bonds, and the 12.5 to 13.0 µm to the oscillations of As–O and Ge–O bonds, which is confirmed by other authors' data.[146,150,155] In the work by Gerasimenko the presence of the band 11.1 µm in $GeSe_x$ glasses is noted.[156] Besides, in this range of the spectrum were the bands 9.0 and 9.4 µm[150] and 9.2 and 9.8 µm;[154] their presence is explained by the presence of SiO_2 in the glass.

Bands having their origin due in carbon are the least known, judging from the number of publications. The 4.94, 6.62, 7.6, and 8.85 µm bands are ascribed to carbon combined with the glass components and the 4.26 to 4.3 µm bands,[150,152,156] to carbon in the form of CO_2.

We now come to the examination of intrinsic absorption bands. According to the data presented in the work by Borisevich and others,[149] in sulfide glasses the 7.60 and 10.15 µm bands arise from oscillations of S–S bonds and 10.1 µm of As–S bonds.

In selenide glasses apart from those discovered by us and confirmed in the series of works bands, 13.5 to 13.6 µm from Se–Se bonds in chains, 14.2 to 14.5 and 20.0 to 21 µm from As–Se bond, to selenium bonds, the bands 10.5,[155] 11.9,[156] 36.0 and 42.0 µm[157] are also ascribed.

We ascribed the 17.8 to 18.0 µm band, which wasn't found in other works, to the oscillations of Ge–Se bonds. In the work by Hilton the bands 21.0 and 42.0 µm[158] are ascribed to these bands, but taking into account that oscillation frequencies of germanium bonds are higher than those of arsenic ones, it appears more correct to ascribe the 17.8 to 18.0 µm band to Ge–Se and the 21.0 µm band to As–Se bonds as was done, for example, in Moynihan's work.[153]

The systematization performed by us of the alignment of interpreted extrinsic absorption bands in the spectrum (Table 5), based on the analysis of 16 research works where the most complete data are presented, first and foremost shows that in some cases the difference exists in band interpretation, put forward by different authors. However, regardless of the interpretation, all these bands were revealed, and they occupy the major part of the spectrum.

It is obvious that to apply ChG in devices it is necessary to have raw materials, as free of oxygen, hydrogen, and carbon as possible and to carry out glass synthesis in conditions which preserve their properties or to clean the glass after synthesis by some method. Such methods are known and they will be taken up in Chapters 3 and 4.

Between the groups of extrinsic bands there are small spectral ranges with neither bonds nor their overtones, hence the glasses can possess minimum absorption there, namely:

1. For sulfide glasses: 2.15 to 2.25, 2.35 to 2.5, 3.15 to 3.54, 4.1 to 6.2, and 8.0 to 8.6 µm
2. For selenide glasses with arsenic: 2.0 to 2.27, 2.35 to 2.70, 3.12 to 3.44, 5.0 to 6.25, 6.67 to 7.14, 8.0 to 8.5, and 11.5 to 11.8 µm
3. For selenide glasses with germanium without arsenic: up to 6 µm — the same, then 7.2 to 7.7, 8.2 to 10.6, and 11.9 to 12.4 µm

TABLE 5

The Location of Extrinsic and Intrinsic Absorption Bands in ChG Spectra

γ (cm^{-1})	λ (μm)	Bands position	Ref.	γ (cm^{-1})	λ (μm)	Bands position	Ref.
5000	2	S–H	147, 148		8.9	As–O	153
	2.05	S–H	147, 148		9	Si–O	150
	2.1	S–H	150		9.1	As–O	146, 154
4500	2.29	H$_2$O	147, 148, 150		9.2	Si–O	154
	2.3	Se–H	147, 148		9.4	Si–O	150
	2.32	H$_2$O	147, 148		9.5	As–O	146, 153, our
4000	2.55	S–H	147, 148, 150		9.6	As–O	Our
	2.73	OH$^-$	86, 147, 148, 156, our		9.8	Si–O	154
	2.78	OH$^-$	147, 148, 153, our	1000	10.1	As–S	149
	2.84	OH$^-$	147, 148, 153, our		10.2	As–O	154
3500	2.9	OH$^-$	146–148, our		10.2	S–S	149
	2.92	OH$^-$	153, 156, our		10.4	As–O	153
3000	3.09	S–H,	147–150, 159		10.5	Se–Se	155
	3.09	S–OH$^-$			10.7	Se–O	152
	3.45	Se–H	151		10.7	As–O	154
	3.55	Se–H	147, 148, 153		10.7	H$_2$O	147, 148
	3.64	S–H	147–149	900	11.1	Ge–O	156
	3.69	S–H	147, 148		11.4	As–Se	153
2500	4	S–H	147–149		11.9	Se–Se	156
	4.1	S–H	146–148, 150	800	12.5	As–O	86, 154, our
	4.1	Se–H	153		12.5	Ge–O	155, our
	4.15	Se–As–H	147, 148		·12.7	As–O	153, our
	4.26	CO$_2$	150, 152, 156		12.7	Ge–O	86, 146, 150, our
	4.3	Se–H	159		13	As–O	146, 158, our
	4.45	Se–H	146, 158, our		13	Ge–O	Our
	4.5	Se–H	146, our		13.5	Se–Se	152, 155, our
	4.57	Se–H	86, 153, our		13.6	Te–O	146
	4.7	Se–H	147, 148, our		13.6	Se–Se	Our
	4.9	Se–H	Our		14.1	As–O	86
	4.94	Ge–H	159	700	14.4	Se–H	151
	4.94	C–S	150		14.4	Se–Se	Our
2000	5	Te–H	146		14.4	Te–O	146
	6.3	H$_2$O	86, 146, 152, 153,		14.5	Se–Se	153, our
	6.3		156, our		14.8	Sn–O	Our

TABLE 5 (continued)

The Location of Extrinsic and Intrinsic Absorption Bands in ChG Spectra

γ (cm^{-1})	λ (μm)	Bands position	Ref.	γ (cm^{-1})	λ (μm)	Bands position	Ref.
1500	6.62	CS$_2$	159		15	Sn–O	Our
	7.17	Se–H	151		15.9	Se–H	153
	7.5	As–O	153		16	Se–O	158
	7.6	S–S	149	600	17.8	Ge–Se	Our
	7.6	C–S	152		18	Ge–Se	Our
	7.8	Se–H	153	500	20	As–Se, As–O	Our
	7.9	As–O	150, 153		20.3	As–Se, As–O	Our
	7.9	Ge–O	Our		20.8	As–Se	147, 148,
	8	Ge–O	Our		20.8	As–Se	152, 153, 155, our
	8.1	Ge–O	146		21	Ge–O	158
	8.25	P–O	146		21	As–Se, As–O	Our
	8.6	As–O	154		29	As–Se	153
	8.7	As–O	146		36	Ge–Se	157
	8.85	C	152				

2.2   OPTIC CONSTANTS OF CHALCOGENIDE GLASSES[21]

In 1964 to 1965 Ajio and Yefimov were carrying out systematic investigations of changes in the refractive index and dispersion with composition of glasses of the following systems: As–Se, Ge–Se, As–Ge–Se, As–Ge–S–Se, As–Ge–Se–Sn, As–Ge–Se–Pb, As–Ge–Se–Sb, As–Ge–Se–Te, As–Ge–S–Se–Sn, and As–Ge–Te–Tl.

The complex glasses were produced from the principal system As–Ge–Se at the successive replacement of the components by their analogues in the periodic system, i.e., Ge → Sn → Pb, As → Sb, Se → S → Te. Besides, a series of glasses was synthesized containing Sb + Sn, Sn + Pb, Pb + Te, and other combinations of elements. To compare the influence of one or another component of the glass on the value of refractive index in complex systems, the glasses were investigated with the constant content of arsenic 20 at.% or equal contents of arsenic and germanium. The weight of simultaneously synthesized glasses was equal to 100 to 150 g.

Time-temperature synthesis regime was worked out for each system, respectively. After the molling the control specimens were thoroughly checked for monophase character and after that their spectral transparency was measured. Only pure glasses, free of extrinsic bands, were used for manufacturing prisms. In prisms for the measurement of the refractive index the dimensions of large surfaces were 35×35 mm, the refracting angle $11° \pm 30'$, and the refracting edge, perpendicular to the foot — accurate within $10'$. Large surfaces were polished with a precision of 1/2 of the interference band; the foot and the third small side remained ground. Over one of the large burnished sides (cathetus) a specular layer of aluminum was applied.

The measurements were carried out using an IR goniometer IR24 with a precision of ±0.0003. The reproducibility of results for different meltings was $±1 \times 10^{-3}$. The wavelength range investigated is located in proximity of the principal electron absorption band; that's why only a monotonous decrease of the slope in the curve n-λ was observed here. Table 6 shows examples of refractive index values for glasses of various systems. The measurements were conducted in the 1- to 11-μm spectrum range at intervals of 0.4 to 0.8 μm. It is evident that in most cases the refractive index dispersion varies with composition to a much lesser extent than the absolute values of refractive index. Therefore, in the analysis of the refractive index of the glass composition dependence, we take the value for wavelength 1.8 to 2.2 μm where the dispersion of the refractive index is at the most dispersion. Besides, only in this range is the comparison with oxide optical glasses and the majority of crystals possible. Having in mind that the refractive index is determined not only by the properties of electron shells of atoms, but also by the density of their packing in substance structure, to find out individual peculiarities of different elements and analyze the connection of optic characteristics with glass composition and structure it is necessary to consider the refractive values along with the changes of refractive index.

As is known, the trend is to use for oxide glasses the molecular refraction value:

$$R_m = \left(\frac{n^2 - 1}{n^2 + 2} \right) \frac{\overline{M}}{\rho}$$

and the average molecular glass weight M is calculated at that from the formula

$$\overline{M} = \Sigma \, X_i M_i$$

where X_i and M_i are, respectively, mole fractions and molecular weights of oxides which enter in the composition of glass. It should be noted, however, that the concept of molecular weight and hence of molecular refraction can't be considered strict enough when applied to substances in the vitreous state, because in the network of glass we, as a rule, can't distinguish ultimate molecules. As to the calculation of molecular refractions of ChG, it is impossible because they are synthesized, in all cases, not from the compounds but from the elements taken in arbitrary relations, hence we can't use values of molecular weights. Refraction for ChG can be calculated from the formula:

$$R_a = \frac{n^2 - 1}{n^2 + 2} \frac{\overline{A}}{\rho}$$

where $\overline{A}$ is the average atomic weight of glass.

It will be recalled that the concept of average atomic weight was introduced by us when calculating the total volume concentration of atoms in glass. Application of $\overline{A}$ for refraction calculation enables one to evaluate the average atom's refraction in the given glass R_a, which characterizes the average value of their polarization.

TABLE 6
Refractive Index Values for Chalcogenide Glasses of Various Systems

Composition in at.%			Other components	Refractive index wavelength (μm)								
Ge	As	Se		1	1.8	2.2	3	4.6	6.2	8.6	10.2	11
20	—	80	—	2.4932	2.4345	2.4266	2.419	2.4138	2.4109	2.4058	2.4022	2.4002
10	—	65	S-25	2.3704	2.3173	2.3103	2.3034	2.2984	2.2954	2.2908	2.2869	—
17	—	80	Sn-3	—	2.4501	2.442	2.4345	2.4289	2.4256	2.4214	2.4178	2.4158
—	20	80	—	2.6823	2.6013	2.5906	2.5814	2.5745	2.5718	2.5683	2.5664	2.5657
—	30	50	Te-20	—	—	2.949	2.9324	2.9211	2.9162	2.9109	2.9074	2.9055
—	30	60	Tl-10	—	3.0376	3.0185	3.0018	2.9899	2.984	2.9765	2.9711	2.969
—	30	45	Tl-15, Te-10	—	3.2908	3.2623	3.2369	3.2219	3.2145	3.2062	—	—
12.5	20	67.5	—	2.6626	2.59	2.5803	2.5713	2.565	2.5608	2.5559	2.5523	2.55
20	20	60	—	2.5983	2.5338	2.5251	2.5168	2.5109	2.5069	2.501	2.4961	2.4941
20	20	30	S-30	2.4096	2.3629	2.3564	2.3501	2.3439	2.3389	—	—	—
22	22	36	Te-20	—	2.8282	2.8129	2.7989	2.7905	2.7861	2.7807	2.7769	2.7745
17	20	58	Sn-5	2.7079	2.6318	2.6214	2.6114	2.6045	2.6002	2.5941	2.5892	2.587
17	20	28	Sn-5, S-30	2.5312	2.4744	2.4665	2.4586	2.4517	2.4462	2.4366	—	—
25	20	50	Pb-5	—	2.7421	2.7299	2.7185	2.71	2.705	2.699	2.6936	2.691
12.5	10	67.5	Sb-10	2.7606	2.6748	2.6637	2.6529	2.6452	2.6407	2.6348	2.6303	2.6272
12.5	—	67.5	Sb-20	2.8906	2.7888	2.7755	2.7645	2.7547	2.7494	2.7425	2.7368	2.7336
5	20	70	S-5	—	2.5379	2.5286	2.5201	2.5139	2.5109	2.507	2.5042	2.5021

From the accuracy of the measurement of refractive index and the value of maximum density spread for different experiments the possible error in the refraction calculation should be estimated at $\pm 1 \times 10^{-2}$. Refraction was calculated for wavelength 1.8 μm; dispersion coefficient — for wavelength 2 μm.

We consider, first, the dependence of optic characteristics from the composition of the glasses As–Se, Ge–Se, and As–Ge–Se and then establish what alterations are introduced by complete or partial replacement of any element by its analogues in the periodic system. In Figure 37 (a and b) four characteristics are presented: dispersion coefficient $V_{2.0}$, refraction $R_{1.8}$, concentration of atoms in 1 cm³ of glass, and refractive index $n_{2.0}$. It is seen that in the glasses of the As–Se system (Figure 37a) refractive index and concentration of atoms enhance rapidly as the content of arsenic increases; meanwhile refraction remains practically constant. Hence the refractive index enhances here, influenced by the increase of atomic concentration, resulting from the decrease in amount of selenium chain structures.

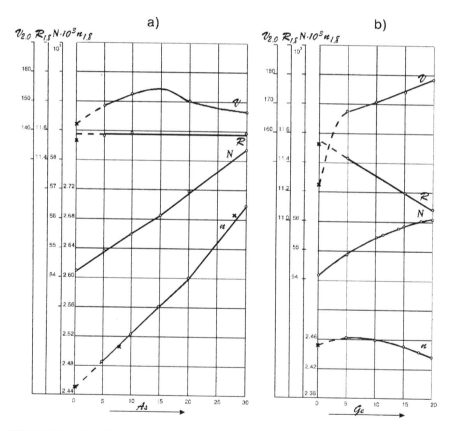

FIGURE 37. The change of atomic concentration (N), dispersion coefficient (v), refraction (R), and refractive index (n) in the systems (a) AsSe and (b) GeSe.

In the Ge–Se system as the content of germanium is increasing, the concentration of atoms in all glasses increases as in the As–Se system. The refraction, however, decreases linearly and, beginning with 10 at.% Ge it prevails over the influence of concentration increase. Hence it follows that initial insertion of germanium doesn't essentially change the refractive index and after that it falls from 2.45 to 2.43. The dispersion coefficient grows as the content of germanium increases, ranging up to about 180 (Figure 37b).

It has been demonstrated above that in the As–Ge–Se system there exist four regions of compositions with different structures. Based upon the glass $As_{20}Se_{80}$ and to replace selenium successfully with germanium, the course of curves of atomic concentration and refraction in the region of selenium chain structures remains similar to that in the Ge–Se system. As a result the refractive index decreases only slightly. In the region from 10 to 14 at.% Ge, where there are no chain structures and only the replacement of double selenium bridges with the single ones takes place, the atomic concentration ceases to increase and the refraction keeps decreasing, which leads to a more detectable decrease of the refractive index. This decrease becomes especially considerable after the pseudobinary cut $AsSe_{3/2}$–$GeSe_{4/2}$, where the appearance of As–As bonds leads to disintegration of the structure — the minimum of atomic concentration lies in this region (Figure 38a).

Both the refraction and concentration of atoms start to be built up in the region of compositions with great lack of selenium for enrichment of valent bonds of arsenic and germanium, where the elements of the structures As–Ge and $AsGe_2$ appear. This, naturally, causes the enhancement of the refractive index ranging up to 2.7590. The dispersion coefficient passes the maximum in that very region of composition, where for the remaining characteristics examined the minimum is observed. Therewith the dispersion coefficient of glasses with the excess of selenium, in comparison with the glasses of the binary systems, has intermediate values and with the lack of selenium it becomes much lower than even in the As–Se system (the minimum value 130).

Let us now examine the optic characteristics of the glasses with additions of different elements of the periodic system; therewith we will assess the influence of additions through the value of the change of refractive index and dispersion coefficient at the rate of 1 at.% of an inserted element.

2.2.1 SULFUR

Sulfur was added to all glasses of the initial system as As_2S_3 in a constant amount equal to 30 at.%. According to the value of the ordinal number, sulfur is the only one among the elements investigated by us which causes the decrease of the refractive index and refraction of glasses of the initial system and the increase of dispersion coefficient. The general form of curves remains intact, which confirms the identity between structural elements $AsS_{3/2}$ and $AsSe_{3/2}$. As we can see from Figure 38b, the concentration of atoms in sulfur-containing glasses is higher than in pure selenium ones, hence the resulting reduction of the refractive index, caused by more sufficient role of the refraction decrease, i.e., by the lower electron polarization of sulfur atoms in comparison with selenium.

It should be noted that the sulfur-containing glasses are characterized by the largest values of dispersion coefficient (180 to 212) in the investigated part of the spectrum.

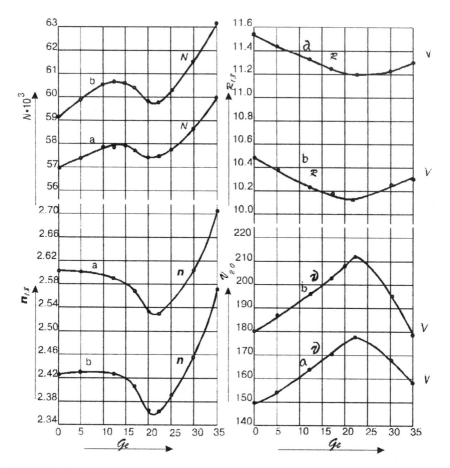

FIGURE 38. Comparison of glass composition dependence of optical characteristics and the number of atoms in (a) AsGeSe and (b) AsGeSeS systems (As cut = const 20 at.%, S = const 30 at.%).

2.2.2 TELLURIUM

Tellurium was added to the glasses instead of selenium in the amount of 10 and 20 at.% along the cut of the principal system at.% As = at.% Ge. On the addition of tellurium considerable increase of the refractive index and glasses refraction is observed with a simultaneous decrease of dispersion coefficient and atomic concentration.

The general form for the curves of the refractive index change and atomic concentration in tellurium-containing glasses is identical and, in comparison with As–Ge–Se glasses, is characterized by the successive disappearing of bending in the region close to the pseudobinary cut. This fact and also a considerable decrease of atomic concentration are evident of the general disintegration of the structure at the expense of tellurium atom's sizes. This circumstance slightly reduces the effect of the refractive index increase caused by the enhancement of the electron polarization of atoms; however, its growth is, nevertheless, considerable and averages $(+126 \pm 26) \times 10^{-4}$ per 1 at.% Te.

2.2.3 ANTIMONY

Antimony was inserted into the glass $As_{20}Ge_{12.5}Se_{67.5}$ at the expense of arsenic in the amount of 5, 10, 15, and 20 at.%. The change of the refractive index on addition of antimony averages $(+91 \pm 13) \times 10^{-4}$ per 1 at.% Sb, of refraction $(+4.8 \pm 0.1) \times 10^{-2}$ and dispersion coefficient (-1.4 ± 0.2). However, as in the glasses with tellurium, the decrease of atomic concentration influences the growth of the refractive index.

2.2.4 TIN

The content of tin in all glasses investigated amounted to 5 at.%, but in this amount tin also has a pronounced effect on the optic characteristics. Therewith the form of curves of the refractive index and atomic concentration remains identical to the glasses As–Ge–Se (Figure 39). However, if in the principal system in the region with the lack of selenium both characteristics keep decreasing in tin-containing glasses, extreme points of the curves "composition-property" are on the line $AsSe_{3/2}$–$(Ge,Sn)Se_{4/2}$, which establishes a line of demarcation between regions with excess and lack of selenium. This confirms the previously applied considerations about the appearance of trigonal structural elements $SnSe_{2/2}Se_{1/3}$ in the region with the lack of selenium. Hence, tin occurs in ChG in two structural forms; each of them introduces its own change in the optical characteristics. As follows from Figure 39, exactly in the region with the lack of selenium the greatest refractive index increase is observed, because in this case, apart from those examined above, two effects take place: increases of atomic concentration and refraction. According to the above, the dispersion coefficient grows slightly at first, then decreases.

2.2.5 LEAD

The addition of lead into glasses of the majority of compositions of the As–Ge–Se system, as it was shown, leads to the appearance of the second phase in the form of the finely divided crystals PbSe. That's why we could evaluate the influence of lead on the optic characteristics of glasses only by the example of individual compositions in the region with the minimum content of selenium. Replacing germanium with lead in the composition $As_{20}Ge_{30}Se_{50}$ caused rather sufficient growth of the refractive index (about $+300 \times 10^{-4}$ per 1 at.% Pb), because apart from the greater electron polarizability of a lead atom in comparison with germanium, there exists the influence of the effect of atomic concentration increase, as in the region with the lack of selenium, lead, like tin, forms more dense trigonal nodes $PbSe_{2/2}Se_{1/2}$.

2.2.6 THALLIUM AND TELLURIUM

On the insertion of thallium in tellurium-containing glasses the values of atomic concentration reach the maximum in the region of compositions where the amounts of tellurium and thallium are in the ratio from 2:1 to 1:2. Therewith the content of

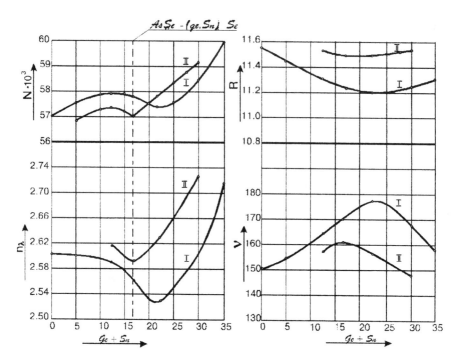

FIGURE 39. Comparison of optical characteristics and atomic concentration dependence on the glass composition in the systems: (I) As–Ge–Se and (II) As–Ge–Sn–Se (Sn 5 at.%) As cut = const 20 at.%.

each of the elements may be brought up to 20 at.%. No wonder that these glasses appeared to be the most highly refractive among all those we had investigated, and their dispersion coefficient the lowest.

In Table 7 are presented values, the results of this work, of the alterations of optical constants, introduced by various elements, with regard to the tolerable content of these elements in ChG. We also managed to plot an Abbe type diagram in the coordinates $n_{2.0}$ to $v_{2.0}$ (Figure 40) for all the studied glasses. It can be seen that ChG are located on the diagram like the glasses from the optical catalogue, i.e., with the increase of the refractive index the dispersion coefficient decreases.

2.3   THERMAL CHANGE IN REFRACTIVE INDEX[22,23]

The change of refractive index with temperature for devices operating in a broad temperature interval degrades the parameters of the optical system. That's why for application of ChG in lens optics information about their thermooptical properties is necessary.

The changes of thermal change of the refractive index with composition were determined for both glasses of elementary systems As–Se, Ge–Se, and As–Ge–Se, and complex glasses, formed on insertion of elements of the IV group — Sn, Pb; V group — Sb, Bi; VI group — S, Te, and also Tl and I, into the principal compositions.

TABLE 7

Variations of Optical Constants When the As–Ge–Se Glasses Components Are Substituted by Their Periodic System Counterparts

Type of substitution	Increment per 1 at.% of the component introduced		Maximum content of the component introduced (at.%)	Variation of optical constants in the condition of maximum content of the component introduced	
	$\Delta n_{1.8} \cdot 10^4$	$\Delta \gamma_{2.0}$		$\Delta n_{1.8}$	$\Delta \gamma_{2.0}$
Ge → Sn when excess of Se	+56 ± 8	−1.9 ± 0.5	10	+0.06 ± 0.01	−19 ± 5
Ge → Sn when shortage of Se	+226 ± 21	−4.7 ± 0.5	10	+0.23 ± 0.02	−47 ± 5
Ge → Pb	+336 ± 60	−5.4 ± 0.2	10	+0.34 ± 0.06	−54 ± 2
As → Sb	+91 ± 13	−1.4 ± 0.2	20	+0.18 ± 0.03	−28 ± 2
Se → S	−54 ± 5	+1.0 ± 0.3	60	−0.32 ± 0.03	+60 ± 18
Se → Te	+126 ± 26	−2.3 ± 0.5	30	+0.38 ± 0.08	+70 ± 15

The measurements of the refractive index at different temperatures in the range 20 to 120°C were carried out at a special goniometric installation according to procedures which allows one to get the value of the thermal change of the refractive index in the wavelength region 2 to 7 μm with the precision of $\pm 3 \times 10^{-4}$. According to results of measurements for the given wavelength the coefficient of thermal change β was calculated as an average change of the refractive index at 1°C temperature rise in the interval 20 to 120°C. β deviations from the average values were not more than $\pm 3 \times 10^{-6}$.

Systematical investigations of the glasses As–Se, Ge–Se, and As–Ge–Se have shown that there are compositions with negative values of thermooptical constant. Sulfur and iodine reduce thermooptical constant of the principal systems, and antimony, bismuth, tellurium, and thallium enhance it, therewith the last influence it more than others. Tin and lead behave differently, depending on the structure of the glasses in which they are inserted.

In 1967 Hilton and Jones demonstrated that the thermal change of the refractive index in ChG, depending on the composition, can be explained with the help of the approximate Lorenz-Lorenz formula:

$$1/n^3 \left(\Delta n / \Delta T \right) = 1/6 \left[1/R \times \left(\Delta R / \Delta T \right) - 3 \left(\Delta L / L \Delta T \right) \right]$$

where $\Delta R/\Delta T$ is temperature change of refraction and $\Delta 1/1\Delta T$ is thermal coefficient of linear expansion.[160]

It can be seen that as the temperature change of refraction goes up, the value β must increase; with the growth of CTE, it must decrease. The analysis of the

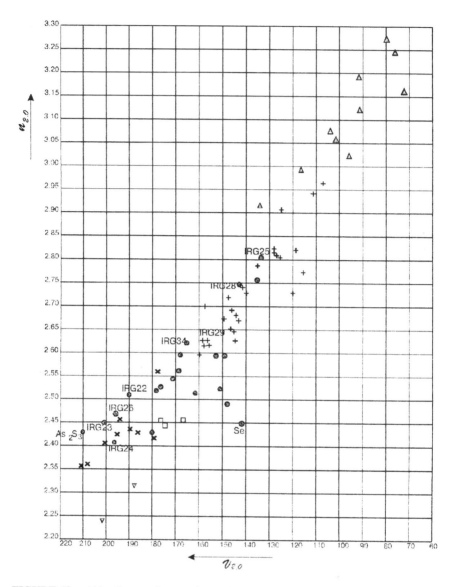

FIGURE 40. Abbe diagram for experimental and commercial glasses in coordinates $n_{2.0}$ to $v_{2.0}$: ∇ — Ge–S–Se; $\square$ — Ge–Se; $\bullet$ — As–Se; As–Ge–Se and commercial glasses + — AsGeSe with additions Sn, Pb, Sb; $\triangle$ — As–Ge–Se with additions Te and Tl; $\times$ — glasses based on As_2S_3 and Se.

experiment really shows that β enhances in covalent-correlated systems at equal CTE when the elements with the higher electron polarizability, namely, Sn, Pb, and Te, are inserted into the composition of glasses. In the glasses which contain iono-genic elements Tl and I, where constant dipoles were formed, β decreases due to the lower temperature change of refraction. Negative values of β appear in the glasses with sufficiently large CTE and the low electron polarizability tendencies.

In Figure 41 our data are presented graphically in coordinates proposed by Hilton and Jones, where the values they had received for some monocrystals are also given for comparison. It can be seen that the glasses are intermediate between covalent and ion crystals. Therewith the glasses with spatial covalent structures, formed by the elements of the groups 4, 5, 6 — As, Ge, Sb, Bi, Se, Te — are located closer to covalent crystals in the region of positive values of β, whereas the glasses with chain structure, i.e., with the greater content of sulfur and selenium, and also the glasses with iodine and thallium adjacent to ion crystals, their β are negative. The behavior of Sn and Pb is determined by the change of their coordinate numbers and hence of structural forms — at transfer from the region of the compositions with excess of chalcogen to the region with its lack. In this connection electron polarizability of atoms, hence the refractive index influence upon the coefficient of thermal change, gets changed.

As a result of these investigations, the possibility of manufacturing ChG with β = 0 for optical systems was established.

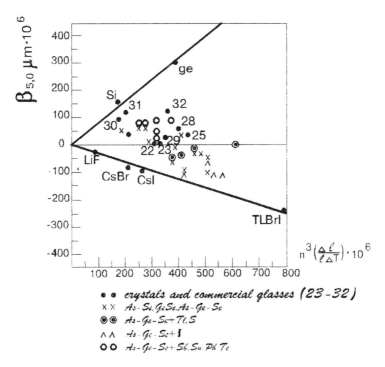

FIGURE 41. Coefficient of temperature change of the refractive index at wavelength 5 μm as the function of refractive index and CTE.

3   Elaboration of Commercial Glasses

3.1   OPTICAL MATERIALS FOR THE INFRARED RANGE

Modern infrared instruments impose heavy demands on the materials that serve as windows, filters, and details of optical systems in devices for thermal vision, self-guidance, IR photographing, etc. As is noted in the monograph by Lloyd, ideally, apart from high transparency in IR-spectral range, their refractive indices must be high, independent of temperature, zero dispersion, zero coefficient of thermal expansion, high surface hardness, mechanical strength and chemical stability, and compatibility with covers.[217]

In spite of the large amount of substances transparent in IR-spectral range, there is no commercial material which at least partly answers all these demands and its elaboration is extremely difficult. That is why it is so important for manufacturers of IR instruments to know the potentialities of existing materials so as to apply them with maximum benefit.

We begin by analyzing the properties of IR mediums which are used most extensively and determining the place of chalcogenide glasses among them. In compliance with the requirements of IR equipment, all the existing materials may be combined into three groups according to transparency:

1. Transparent up to 3 μm
2. Transparent up to 5 μm
3. Transparent up to 14 μm and beyond

This correlates approximately with the positions of the first, second, and third atmospheric windows. According to the evidence from Kruse and others, the first atmospheric window is located in the range from 0.75 to 2.5 μm (near-infrared), another one from 3.0 to 5.0 μm (middle infrared), and a third from 7.5 to 14.0 μm (far infrared).[161]

The first group of materials is the most numerous one. Here are all transparent crystals, many optical glasses with water absorption bands removed, and also quartz glass and devitrified glass such as Pyroceram, which satisfies the requirements by their heat resistance and strength.

The second group is much smaller. Here remain such crystals as lithium fluoride, which can't be used everywhere due to the water absorption bands, which are difficult to remove, and insufficient chemical stability; and fluorite — very transparent up to 8 mm but with high α at a comparatively large value of Young's modulus, which doesn't allow one to use it substantially at great temperature differences. Among monocrystals BaF_2 should also be noted: its transparency range is wider, but its chemical stability is lower than in fluorite; and also a series of crystals such as CdF_2, PbF_2, Al_2O_3, MgO, TiO_2, and others: their major drawback is the considerable difficulties in manufacturing specimens of commercial sizes.

In spite of the drawbacks listed above, all these monocrystals can be used and are used, in fact, in installations that don't require large sizes and great temperature differences. Into this group of substances fall sulfur-containing ChG including the material, which was the first to appear in trade catalogues — trisulfide arsenic, called in the U.S. "Servofrax". This glass is widely used in spite of the comparatively low T_{ds} (212°C) and quite high α which, however, differs only slightly from that of fluorite.

The third group is the smallest one.

Germanium and silicon are transparent beyond 14 μm. True, they both have high absorption by free acceptors and, hence, considerable radiation at elevated temperatures; however, in the objectives of optical systems Ge and Si crystals are quite widely applied.

Halogenides of alkaline metals are very transparent in this region, but their chemical stability is low and attempts to increase it by the application of coatings were of no success.

AgCl is very transparent but a soft crystal, and it is successfully used for riber extension.

Mixed thallium halogenides known as KRS-5 and KRS-6 are further transparent and can be produced with sufficient homogeneity. A considerable toxicity, which manifests itself at synthesis and processing, prevents their wide distribution.

Zinc sulfide and selenide and also cadmium telluride should be noted apart. These are very transparent, heat resistant, and rigid materials, but in the crystalline form they are still grown only up to 40 to 60 mm across.

Pressed polycrystals, so-called "Irtrans" or optical ceramics, are high on the list of IR materials. Actually Irtrans can be extracted from any crystalline substance, if to take it in the form of finely divided powder and to put it through the heat treatment under pressure. The material preserves its optical and physicochemical properties, but its dimensions may be larger (up to 200 to 300 mm across). Polycrystalline MgF_2 (Irtran 1), ZnS (Ir2), CaF_2 (Ir3), ZnSe (Ir4), and CdTe (Ir6) were produced commercially. In the last few years ZnSe has been produced in the form of polycrystalline plates 1 cm thick with the area of several m^2, by the method of chemical precipitation from the vapor phase as a result of zinc vapor reaction with hydrogen selenide. It should be noted that this technology is complicated and dangerously explosive. Anyway, optical polycrystals are quite progressive IR materials, especially for illuminators in IR devices on airplanes and rockets. However, as early as in the mid 1970s, American and British companies, already having irtrans at their disposal, had elaborated and begun to produce ChG.

Fragibility and comparatively low hardness and thermal stability are the most essential drawbacks of vitreous materials. An additional drawback for ChG is lower heat resistance: Savage and Nielsen, based on their own investigations and taking into account the value of binding energy of sulfides and selenides, came to the conclusion about the impossibility to obtain ChG at the T_{ds} higher than 500°C,[84] and our experience also confirms that. However, the maximum temperature for utilization of oxygen-free glasses is determined not only by the T_{ds}, but also by the furnace life of an article at high temperatures. We established the period of possible exploitation on the specimens of two glasses with the T_{ds} equal to 270 and 440°C in the furnace of an impression viscosimeter. The cold specimen was put into the furnace which had been heated beforehand, and the point of time after which the viscosity of the specimen reached 10^{10} P was fixed. One can see (Figure 42) that the first glass in the course of 60 s withstands temperature of 420°C, and the second at 800°C. Therewith neither alteration of the shape of the article nor transmission decrease was observed. This time interval is quite sufficient for some operating conditions. Then, despite the comparatively large values of CTE (from 40 to 10 · 10^{-6}°C^{-1}) thermal stability of these glasses is quite satisfactory. Recall that the product of CTE and the elastic modulus enters into the formula which determines the ability of the material to be resistant to thermal actions: $\Delta T = (\sigma_{tension}/\alpha \times E)\sqrt{a}$. In ChG the values of the elastic modulus are four times greater than in oxide glasses and up to ten times smaller than in optical crystals, which fully compensates the difference in CTE.

The general beneficial property of ChG is their high resistance to the influence of damp atmosphere and acid water solutions (chemical stability of all optical glasses is determined by these very parameters) and also high resistance to hard radiation.

The totality of all these characteristics leads to a wide application of ChG as components of optical devices and also as cuvettes for the investigation of aggressive liquids. ChG can be combined in two groups by their transparency:

1. Glasses transparent up to 9 to 11 μm
2. Glasses transparent up to 14 to 18 μm

The glasses of the first group may be used in the first and second atmospheric windows (1.5 to 2.7 and 3.5 to 5.5 μm, respectively). Their transmission limit derives from sulfur presence in the compositions. Within the group the glasses differ mainly in their thermal and mechanical properties. Their optical constants change only slightly with composition.

The glasses of the second group are transparent in the third atmospheric window (8 to 14 μm), the most beneficial one for IR optics. Selenium or selenium with tellurium and sometimes with small sulfur addings serves as glass-formers here. In this group are also glasses containing halogens. The considerable changes in the refractive index enable one to produce commercial glasses with quite different optical constants. It should be noted that nowadays in IR-objective manufacturing ChG are mainly used combined with germanium or optical ceramics, mostly with zinc selenide. However, the glasses we had elaborated on in recent years allow us to construct wide-angle apochromatic objectives not using crystalline media.

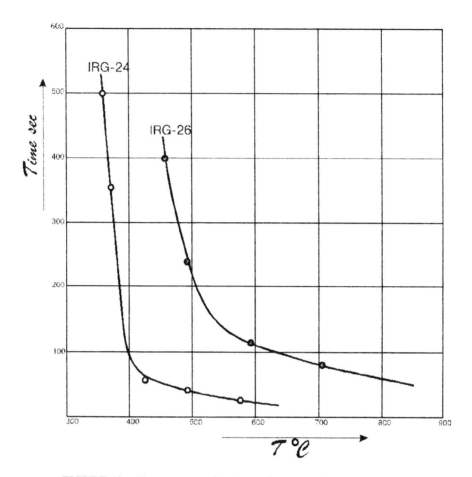

FIGURE 42. Time necessary for glass softening at given temperatures.

3.2 GENERAL PRINCIPLES OF COMMERCIAL GLASS ELABORATION[24]

We carry out the elaboration of glasses that are feasible for commercial manufacturing based on a certain principle: from the investigation of glass-forming and properties of glasses of various elementary systems, the compositions with the best and most beneficial optical and physicochemical properties are chosen and then modified to attain the technological characteristics which enable commercial syntheses. Therewith principal attention is paid to the crystallization tendency in glasses.

One of the most frequent causes for commercial failure lies in the crystallization of glasses. The appearance of crystalline insertions in glass causes light scattering, i.e., it increases optical losses and leads to distortion of the picture and appearance of high stresses which can't be relieved by annealing. Especially frequently observed are the cases of crystallization during manufacturing of optical glass, because the necessity to produce the glasses with certain optical constants hampers alteration in composition for decrease of the crystallizability. That's why the control of crystallization assumes a special significance in developing the formula of new glasses with designated optical and physicochemical properties.

The results of the long-standing investigation of optical glass crystallization are presented in the monograph by Mukhin and Gutkina.[162] The technique they have proposed is still the best and it is widely used when developing and manufacturing optical glass.

The essence of the method is as follows. The glass under study is melted in a platinum crucible at the temperature higher than that of liquidus and then, as quickly as possible, it is poured into the boat made of platinum plate to produce the specimen 8×8 mm at section and 200 mm long. When using platinum boats, it is necessary to take into account the possibility of their interaction with some glass compositions. In such cases ceramic or quartz boats are used. The boat with glass is put into the furnace with the stated and measured temperature gradient. The temperature distribution in the furnace is evaluated as a rule before each experiment. The crystallizability is usually determined at two periods: 6 and 24 h. After the period in the furnace the boat is taken out, quickly cooled in the air, and the specimen is put under study. To determine the crystallization degree and to investigate crystalline phases, the axial sections are prepared out of the specimen in the form of flat, parallel, polished plates 3 to 6 mm thick. In the case when the preparation of axial plates is impossible, for example, because of the specimens' cracking, the cross-sectional ones are prepared at 1 to 3 cm intervals.

In the plates one can observe the crystallization in the thickness of glass, crystal optical peculiarities of separating out crystalline phases, and also the change of crystallization degree along the whole length of the plate. The results are plotted on the diagram of crystallizability. The temperature is plotted along the horizontal axis, the thickness of a specimen in 2:1 scale along the vertical one with the marker of the crystallized part of a specimen in the same scale. The thickness of the layer of the surface crystallization is determined for both upper and lower surfaces of a specimen, which is especially important for the glasses with selective volatility of components. Symbolizations of different degrees of crystallization are given in Table 8.

To evaluate the crystallizability of ChG, including crystallization tendency in the process of synthesis, taking into account the necessity to observe safety measures, we have changed the above-described technique slightly. A certain specimen from the charge of the glass under study is placed into a quartz ampul similar to those used at the synthesis of glasses, but with Ø15 mm and the length corresponding to the size of the crystallization boat. The glass is melted in the usual regime and quenched from the clarification temperature to room temperature, with the ampul kept in horizontal position. Then the ampul with glass is put into the crystallizing furnace, left there for some time, and broken at cooling. Crystallization is evaluated while viewing it through the IR microscope, and when there are concerns the glasses with the transparency edge further 1.2 μm, by analysis of the transmission spectra. The results are plotted according to the accepted symbolization.

TABLE 8
Conventional Symbols for the Polythermal Method
for Crystallizability Determination[162]

Degree	Crystallization type	Thickness of crystalline layer (mm)	Conventional representation
0	No crystalline generation on the glass surface	0.	————
1	Appearance of the first symptoms of the surface changes (folds, small crystalline generation)	0	----------
	Crystalline film, well visible when viewed from the surface, but with a bad end view	<0.1	ＭＭＭＭＭ
2	Crystalline layer	0.1–1	▬▬▬▬▬▯▯▯▯▯▯ Shading height from 0–2 mm
3	Crystalline crust	1–3	▬▬▬▬▬▬▬ Blackened region height from 2–6 mm
	Crystallization from 3 mm to the complete one	3–4	▬▬▬▬▬▬▬ Blackened region height from 6–8 mm

Note: Discrete crystals inside the glass are marked by points, the location of which indicates the location of crystals in the glass.

As a rule, ChG possess the region of partial or even complete crystallization which can't be eliminated by the change of composition without undesirable change of properties. Mukhin and Gutkina offered a method for reduction of crystallizability of optical glasses, which was based on the following rule:

"In the glass-forming systems within the limits of crystallization field of the given compound (in a general way for the compound which melts congruently), the glass corresponding to the composition of the compound will have the maximum crystallizability. For the glasses of other compositions in the same crystallization field the reduction of crystallizability is observed as their compositions are getting more different from the composition of the compound. Crystallizability achieves its minimum in the regions of the given compound's joint crystallization with compounds of other compositions (p. 186)."[162] Hence, we know the equilibrium diagram of the given system which must serve as a basis for developing glass with prescribed properties, it is advantageous to proceed from the eutectics' composition and to perform all necessary changes in the glass composition along eutectic lines or eutectic surfaces.

3.2.1 EUTECTIC PROPERTIES OF CHALCOGENIDE SYSTEMS

The investigation of phase diagrams in any system is a rather arduous process and comparatively few of them were known for ChG, even in the 1970s. In particular, the phase diagram of the system As–Ge–Se was investigated by Vinogradova and others in 1968.[163] The authors found two ternary eutectics $As_{22}Ge_{30}Se_{48}$ and $As_{13}Ge_{32}Se_{55}$. Both glasses are crystallization resistant and are highly hard (T_{ds} = 378 and 348°C, respectively). Only in 1973 Orlova and others investigated the phase diagrams of the subsystems Sb_2Se_3–$GeSe_2$ and Sb_2Se_3–$GeSe$.[98] The authors found three binary eutectics, hence one could expect the presence of ternary eutectics as well. Anyway, the possibility to evaluate *a priori* the eutectics' position in the multicomponent vitreous system could allow one to reduce considerably the experimental work on revealing of crystallization-resistant glasses.

Susarev and others have worked out such a method for saline systems with the ion bond type and for the systems with covalent bond types, but from organic substances.[164] As ChG are covalent-correlated nonorganic polymers, it is rather tempting to calculate with its help the eutectic compositions in complex vitreous systems based on chalcogenes.

The method of calculating eutectic properties according to the data on mutual interaction of components is based on the following properties of the systems. In binary heterogeneous systems the conjugate phases generally have different compositions. The sole exception are the points corresponding to the extreme value of equilibrium temperatures at (P = const) in the systems of the type "hard phases-melt". Here the given phase is in equilibrium with the second phase of true composition and the hardly dispersed mixture of hard phases, i.e., eutectics, is formed at crystallization. To insert the third component into the binary eutectics, the flat system turns into the volume one and eutectic points of binary systems give rise to the corresponding folds on its surface. The eutectics of the ternary system correspond to the meeting place of the noted folds and lies inside the triangle, the vertices of which are binary eutectics. The unequal components' interaction causes fold deviations from the secants connecting binary eutectics with the third component. Deviation value depends on one of two components with which the third one forms the system, the least to the laws of ideal mixtures, and that determines the separating action of an inserted component.

It has been demonstrated that in the condensated systems of the type "hard phases-melt", to evaluate separated action of the components, hence the concentration region of ternary eutectic position on the phase diagram, the value can be used which was called by the authors of the method "relative change of components' partial temperatures A_i". A_i is calculated from the formula:

$$A_i = \frac{2.3}{X_3^{i-3}} \, \lg \frac{T_i^o X_i^{i-3}}{T_{eut}^{i-3}}$$

where X_i^{i-3}, X_3^{i-3} is content of the components in binary eutectics $(i-3)$ in mole parts (Figure 3 marks the inserted component); T_i^o, T_{eut}^{i-3} is fusing temperature of the pure component and binary eutectics with its participance by the absolute temperature scale. The less negative value A_i at that corresponds to the more positive deviations from the laws of ideal blends in the properties of solid phase and to directions of deviations of the repeat crystallization lines from the corresponding secants. As each component takes part in two eutectics to each of them, respectively, two at time values A_i are calculated. In this way the concentration triangle is established, inside of which the ternary eutectic is located.

To calculate the composition of ternary eutectic, it is necessary to reveal the folds of the T surface, having the least tendency to deviate from the secants. The expression used for that is:

$$\Pi = \left[\left(A_1^{1-2} + A_2^{1-2} \right) \left(A_1^{1-3} - A_2^{2-3} \right) \right]$$

Numbers 1 and 2 mark the components of the binary system and number 3 an introduced component. The smaller the value Π is, the more stable is the fold and the less is its deviation from the course of the secants. The authors of the method established that for the system of organic substances with predominantly covalent bond type the point of ternary eutectics is also located close to intersection of the secants, corresponding to the two most stable folds of the surface T. It's obvious that in this point the component relationships will be similar to those in binary eutectics where secants originate and the summarized content of components in mole parts is equal to 1. As a result the composition of ternary eutectics can be found through solution of a system of three equations in three unknowns.

$$X_1 / X_2 = X_1^{1-2} / X_2^{1-2}$$

$$\left\{ \; X_3 / X_2 = X_3^{2-3} / X_2^{2-3} \right.$$

$$X_1 + X_2 + X_3 = 1$$

Belykh was the first to apply this method to evaluate the concentration region of ternary eutectics' position in the system Sb–Ge–Se; the literary data for binary systems were initially refined. Characteristics of nonvariant points of binary systems are given in Table 9.

All of the systems are of the eutectic type. In the system $GeSe$–Sb_2Se_3 there exists an incongruently melting compound $Ge_4Sb_2Se_7$. As an example we refer to A_i calculation for the components of the binary system Sb_2Se_3–$GeSe_2$:

$$A_{Sb_2Se_3}^{Sb_2Se_3-GeSe} = 2.3/0.56 \; \lg \; 890 \times 0.44/718 = -1.083$$

$$A_{GeSe_2}^{GeSe_2-GeSe} = -2.3/0.57 \; \lg \; 1013 \times 0.43/852 = -1.178$$

$$A_{GeSe_2}^{GeSe_2-GeSe} < A_{Sb_2Se_3}^{Sb_2Se_3-GeSe}$$

TABLE 9

Characteristic of the Nonvariant Points

Composition of the alloy (mol%)				Character of
GeSe	GeSe$_2$	Sb$_2$Se$_3$	T_{melt}(K)	the point
57	43	—	852	Eutectic
—	55	45	758	Eutectic
56	—	44	718	Eutectic
64	—	36	736	Peritectic

which corresponds to large, positive withdrawals from the law of ideal blends in the system Sb$_2$Se$_3$–GeSe; that's why the fold must deviate from the secant to the binary system Sb$_2$Se$_3$–GeSe.

In a similar manner A$_i$ for other binary systems are calculated. Calculation of A$_{GeSe_2}^{GeSe-Sb_2Se_3}$ is performed with regard to the liquids lines' fracture in peritectic point (the reaction of formation of incongruently fusing compound). All the values obtained at calculation of the concentration regions of ternary eutectics position in the system Sb$_2$Se$_3$–GeSe–GeSe$_2$ are performed on the scheme (Figure 43). The arrows indicate directions of deviations from the secants. Concentration regions of ternary eutectics' positions are vertically cross-hatched, of ternary peritectics' — horizontally. The calculation of ternary eutectics' composition in the system Sb–Ge–Se is given below. For the folds starting in the binary systems Sb$_2$Se$_3$–GeSe$_2$ Sb$_2$Se$_3$–GeSe and GeSe–GeSe$_2$:

$$\Pi_{Sb_2Se_3-GeSe_2} = /(-1.159-0.683)(-1.178+1.083)/ = 0.175$$

$$\Pi_{Sb_2Se_3-GeSe} = /(-1.083-0.632)(-1.159+1.063)/ = 0.161$$

$$\Pi_{GeSe-GeSe_2} = /(-1.063-1.178)(-0.632+0.683)/ = 0.114$$

Correlation between the values $\Pi 0.114 < 0.161 < 0.175$ shows that the closest to the secant's course, i.e., the most stable is the fold originating in binary systems including GeSe$_2$–GeSe, and GeSe–Sb$_2$Se$_3$. In this way, the composition of ternary eutectics is calculated through the solution of the system of the following equations:

$$\frac{X\,Sb_2Se_3}{X\,GeSe} = \frac{X_{Sb_2Se_3}^{Sb_2Se_3-GeSe}}{X\,Sb_2Se_3 - GeSe} = \frac{0.44}{0.56}$$

$$\frac{X\,GeSe_2}{X\,GeSe} = \frac{X_{GeSe_2}^{GeSe_2-GeSe}}{X_{GeSe}^{GeSe_2-GeSe}} = \frac{0.43}{0.57}$$

$$X_{Sb_2Se_3} + X_{GeSe} + X_{GeSe_2} = 1$$

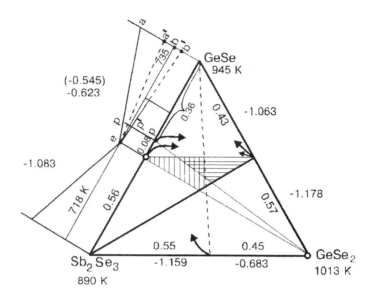

FIGURE 43. Evaluation of concentration regions of location of triple eutectics and peritectics in the system GeSe–Sb$_2$Se$_3$–GeSe$_2$.

We find the component correlation in molar percentage: 39GeSe, 30GeSe$_2$, 31Sb$_2$Se$_3$ or in atomic percentage: Ge$_{21.4}$Sb$_{19.2}$Se$_{59.4}$.

To calculate the fusing temperature of ternary eutectics it is necessary to keep in mind the mutual position of the most infusible component — in the present case GeSe$_2$, the most fusible eutectics (Sb$_2$Se$_3$–GeSe), and the secants which correspond to the most stable fold of the surface T. On this basis the component is chosen by which the calculations are performed.

In the system GeSe–GeSe$_2$–Sb$_2$Se$_3$ the most infusible component doesn't enter the system with peritectic and subtends the most fusible eutectic GeSe–Sb$_2$Se$_3$, and the secant acts as if it separates them. That is why in this case the calculation should be performed with application of the data about the most fusible component, namely, Sb$_2$Se$_3$. The calculation formula is

$$T_{fus}^{eut} = T_i^o - \frac{1-x_i}{1-x^1}\left(T_i^o - T_{eut}^i\right),$$

where T_i^0 and T_{eut}^1 are fusing temperatures of the pure component and more fusible binary eutectics with its participation; X^i and X^1 are content of components in ternary and more fusible binary eutectics in molar parts. Then T_{fus} of ternary eutectics in the system Sb–Ge–Se is equal to:

$$T_{fus}^{eut} = 890 - \frac{1-0.31}{1-0.44}(890-718) = 678° \ K \ \text{or} \ 405°C$$

Correlation of rated and experimental results is given as applied to the system As–Ge–Se. The data for the calculations were taken from the work by Vinogradova and Dembovsky,[163] and in doing so we took ternary eutectics from the triangle $AsGeSe–GeSe_2–As_2Se_3$. According to the phase we can refer the following composition: $As_{22-22.5}Ge_{26-30}Se_{51-47.5}$. Fusing temperatures and compositions of binary eutectics were also taken from the phase diagram of the ternary system. The calculated eutectics has the composition $As_{22.3}Ge_{26.4}Se_{51.3}$. Taking into account the accuracy of the method of experimental determination of eutectic compositions, such agreement may be considered quite satisfactory. We note that the analogous calculation and its experimental test were carried out in 1978 by Turkina for the system Tl–Ge–Se (subsystem $Tl_2Se–TlSe–Tl_4GeSe_4$), also with positive result.[110]

3.3   COMMERCIAL GLASSES TRANSPARENT TO 9 TO 11 μM

Oripigment is the first commercial representative of these glasses. Vitreous As_2S_3 has T_{ds} = 212°C, CTE = 23 × 10^{-6}°C^{-1}, Tg = 153°C, microhardness = 155 kg/mm^2. It is sublimated at 600°C and practically doesn't crystallize. However, it is always more difficult to obtain certain chemical compounds as a homogeneous glass than as a blend. Our aim was to produce a more technologically effective and, if possible, more transparent analogue of arsenic trisulfide without addition of expensive materials. The comparatively cheap natural oripigment was used then as the main component. The glasses like As_2S_3, with improved spectral and technological characteristics, were obtained through the investigation of the system (As_2S_3)–Se. It appeared that small selenium additions hardly changed the properties of As_2S_3, but the resulting glasses were more feasible for commercial manufacturing.

Commercialization of the glass marked IRG23 was mastered in 1961 and it found a wide utility in one of the most mass-produced instruments for thermal vision.

The investigation of the thermal properties of various glasses containing the elements of the V and VI groups of the periodic system had demonstrated that no one composition with T_{ds} much higher than 200°C could be obtained. It is obvious from the information in Chapter 1 that this is connected with the structure, namely, with the covalent correlations of atoms: all the elements of the V group introduce into glasses an equal number of valent bonds with approximately equal energy. It has also been demonstrated above that the only way to increase the covalent correlation of the system, hence, to improve the thermal properties, is in the insertion of elements of the IV group into glass compositions. As silicon doesn't perform stable glasses with sulfur and selenium, we used germanium at the elaboration of commercial materials with enhanced thermal characteristics.

In consequence of the investigation of the crystallizability of a series of glasses in the system (As_2S_3)–Ge–Se, we managed to establish that the most stable glasses are obtained in the region with excess of selenium, where the compositions, suitable for manufacturing, were discovered. The crystallization region for such glasses lies between the values of melting and yield temperatures. The upper limit of crystallization is 550 to 600°C; the lower one 350 to 400°C. There is no total crystallization.

One of such glasses was standardized by the marker IRG24. Its $T_{ds} = 270°C$; CTE = $18.6 \times 10^{-6}°C^{-1}$. The further enhancement of heat resistance and decrease of CTE could be expected only in the region with a great lack of chalcogene, where the energy correlation of sulfur-containing glasses ranges up to 80 kcal/mol, hence one can expect the T_{ds} rise up to 400 to 500°C.

The most heat-resistant glasses are produced along the line As_2S_3–Ge and in the adjacent region. Thus, the glass of the composition $Ge_{30}As_{28}S_{42}$ has $T_{ds} = 470°C$ and its crystallization region lies in the temperature interval 500 to 590°C. However, when attempting to melt such glasses with the weight exceeding 20 g, we faced an insuperable obstacle; on the addition of germanium from the glass melt to oripigment in the process of synthesis, elementary arsenic escapes and precipitates at cooling in the upper part of the ampul in the form of well-shaped crystals, and on its walls in the form of gray crystalline deposit. This process is followed by intense frothing of the melt. This weakly combined arsenic, which easily transfers into the gas phase, creates in the vitrification process pressure on the walls of the container vessels which is equilibrium at the given temperature. When quenching a melt from high temperature, as a result of arsenic vapor condensation on the more quickly cooling walls of the vessel, the balance "liquid-vapor" is disturbed and leads to a further decomposition reaction with arsenic formation. In the case when quenching is performed from the temperature at which the viscosity of glass melt is already high as a consequence of arsenic vapors' escape, the glass melt swelling occurs. The quenching should be carried out at an even lower temperature, where the decomposition reaction stops, but the glasses of the system (As_2S_3)–Ge begin to crystallize at such viscosities. The addition of Sb, Sn, and other elements produced no positive results. We had to replace some amount of oripigment with selenium. In doing so, to remain in the same temperature region, the content of germanium in glasses was enhanced.

As a result, after the addition of excessive selenium to combine escaping arsenic, the high quality glass was produced, without crystals and bubbles. This glass was standardized by the marker IRG26. It has a $T_{ds} = 440°C$ (i.e., it is almost equal to that of heavy flints), and CTE = $11.6 \times 10^{-6}°C^{-1}$. Nevertheless, when melting this glass some arsenic escapes into the gas phase, which can lead to the bursting of containing vessels. For assurance the quantity of the simultaneously found glass IRG26 is limited to 1 kg; hence it can be used in the form of articles not more than 120 mm across and 10 mm thick.

For IR photograph equipment large-size windows are needed as a rule. That is why it was necessary to develop the glass similar to IRG-26 and suitable for the manufacturing of large articles. Along with the system (As_2S_3)–Ge–Se with the great content of As_2S_3, where one can't even hope to find a suitable composition, glasses with T_{ds} higher than 400°C are present only in the principal system As–Ge–Se in the region with a great content of arsenic and germanium. They all, however, have an elevated tendency to crystallization and are unsuitable for commercial production.

To decrease crystallizability and, at the same time, to enhance T_{ds}, some amount of arsenic and selenium was replaced by trisulfide arsenic in stoichiometric regard. Such replacement took place in the glasses of the principal compositions $As_{20}Ge_{30}Se_{50}$ and $As_{30}Ge_{30}Se_{40}$. The experiment has demonstrated that, as it was supposed, equimolecular replacement of As_2Se_3 with oripigment leads to the needed results.

The obtained commercial glass IRG30 has $T_{ds} = 390°C$, CTE $= 12.2 \times 10^{-6}°C^{-1}$, H = 245 kg/mm². It is transparent in the interval 1 to 11.5 µm, i.e., slightly better than IRG24, and can be melted in industrial conditions in the amount up to 20 kg.

3.4 GLASSES FOR OPTICAL SYSTEMS TRANSPARENT TO 14 TO 18 µM

As noted above, the range 8 to 14 µm, due to its special atmosphere properties, is the most beneficial one for the work in the IR spectrum range. However, as with the increase of the radiation wavelength the devices record more and more weakly heated bodies, in some cases; for example, in medicine, more long-wave radiations are used. The materials with maximum transmission in the far-IR region are necessary for that. Possibilities for developing heat-resistant materials decrease here considerably because apart from the lower bond energy of selenium and tellurium in comparison with sulfur, the glasses without germanium are the most transparent ones. That's why peculiar attention was originally paid to the system As–Se.

Recall that the least easily fusible in this system is As_2Se_3 ($T_{ds} = 200°C$) which, due to the enhanced crystallization tendency, is unsuitable for commercial production. One can hope to decrease crystallization preserving at that heat-resistance and transparency, typical of As_2Se_3 only by using tin, lead, and antimony. Investigation of the system As–Sn–Se demonstrated that addition of tin into the region of compositions with excess of selenium really strengthens the glasses, bringing their thermal characteristics to those of As_2Se_3. The upper limit of the crystallization of such glasses is, however, near 500°C and at the molling from the higher temperature the noticeable oxidation of the surface takes place. After 6-h they produce large regions of total crystallization (Figure 44a). The initial crystalline phase is tin diselenide. Small additions of antimony have a beneficial effect on the crystallizability of glasses. Besides, the atomic weight of antimony is close to that of tin and hence antimony can't affect adversely the position of absorption bands. That is why some amount of arsenic was replaced with antimony to decrease crystallization in the glasses of the system As–Sn–Se. To establish optimal content of antimony the crystallizability of glasses was sequentially determined. In Figure 44, b one can see that on the first additions of antimony a slight decrease of crystallization is observed, therewith $SnSe_2$ remains in the initial crystalline phase.

The further insertion of antimony leads to the abrupt enhancement of crystallization. Therewith two different phases appeared in the glass: the high-temperature $SnSe_2$ and the low-temperature one which, judging by the crystals, represented Sb_2Se_3. We managed to produce low-crystallizing glass by further change of the composition (Figure 44, c). The glass obtained was quite suitable for commercial production and it gained the marker IRG25. Its $T_{ds} = 190°C$, and CTE $= 22 \times 10^{-6}°C^{-1}$.

Until recently arsenic has been used as the main component in the majority of commercial ChG. It is quite clear, because glasses of the known arsenic-free systems,

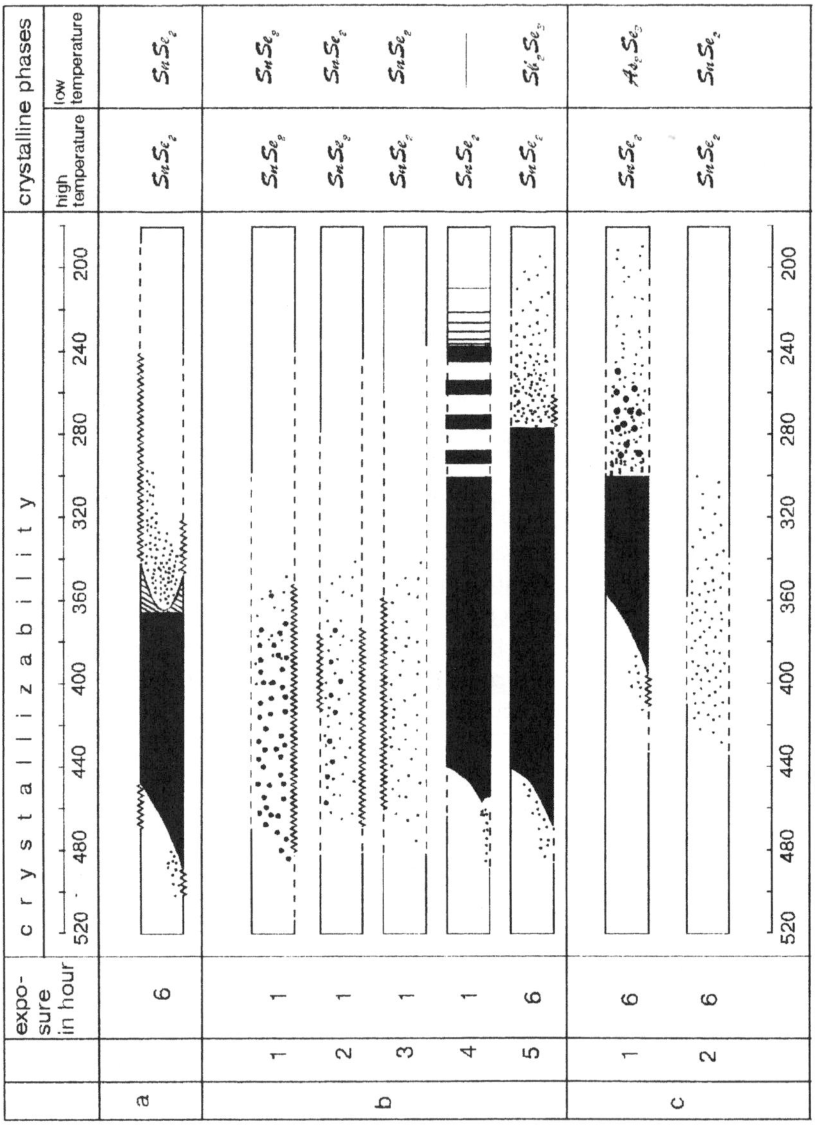

FIGURE 44. Crystallizability and crystalline phases in the glasses of the system As–Sb–Sn–Se.

Sb–Ge–Se, Bi–Ge–Se, and Ge–Se, suitable for application due to their physico-chemical properties, have high crystallizability. The work with arsenic presents, however, certain difficulties deriving from its tendency to oxidation at different stages of synthesis. That's why already in mid 1960s. Kicutzkaya had worked out the arsenic-free commercial glass. Her work was based on results of the investigation of the Ge–Sn–Sb–Se system. As there was no phase diagram, the boundaries of crystallization fields were established as well as at IRG25 elaboration, by the change of the components' relations. The composition of precipitating crystalline phases was evaluated by the crystalline-optical method, and in the case of complete crystallization, by the method of roentgeno-phase analysis (RPhA). In attempts to obtain the glass with high T_{ds} $Sb_5Ge_{25}Se_{70}$ was used as the initial composition. The initial crystalline phase in this melt is $GeSe_2$. Partial replacement of germanium with tin decreases crystallization, but as the content of Sn is enhancing, SnSe separates out, although $GeSe_2$ is still the main phase. The further diminution of crystallizability was achieved due to the decrease of germanium content and insertion of optimal amount of antimony. Therewith covalent correlation of glass network diminished, i.e., T_{ds} slightly lowered, but then the glass was produced which didn't crystallize after 6 h in the gradient furnace. Its $T_{ds} = 200°C$, and CTE $= 22 \times 10^{-6}°C^{-1}$. This glass was standardized under the marker IRG29 and, due to its thermal properties, it can be recommended instead of IRG25 in the wavelength range 1 to 15.5 μm.

3.5   GLASSES FOR CROWN-FLINT PAIRS OF LENS FOR THE OPTICAL SYSTEMS' OBJECTIVES

For the construction of optical devices it is a necessity to eliminate various distortions of the pictures — so-called aberrations. Two principal types of aberration are essential for IR optics. Spherical aberrations can be enhanced through the combination of positive and negative lens made of glasses with different refractive indices. "Correction" of the system by chromatic aberrations is achieved through the combination of the positive lens made of the glass with low dispersion — so-called "optical crown", and the negative one made of glass with much higher dispersion called flint glass or flint. The trend is to name the doublet being produced crown-flint pair. Positive and negative lens of the doublet glasses with the maximum difference in dispersion coefficient are chosen because the correction of other aberrations at that goes more easily due to the smaller curvature of lens. Besides, the higher the refractive index of a flint component is, the greater the optical power in the objective being produced. The peculiarities of light transmission through glass details is discussed at greater length in the collection "Physico-Chemical Foundations of Optical Glass Manufacturing", edited by Dyomkina.[165]

Until recently germanium and optical ceramics have been used in domestic thermovision objectives along with ChG, in particular, with IRG25; however, the application of expensive crystalline mediums in IR devices makes them beyond the reach of a wide range of users. In this regard the pairs made of glass are more promising. In our catalogue there is the glass IRG32, transparent in the region 1.5 to 15 μm: $n_{10.0} = 2.9731$, dispersion coefficient $v_{8-12} = 113$, $T_{ds} = 255°C$,

CTE = 14.7×10^{-6}°C^{-1}. However, this glass didn't gain wide acceptance due to its high crystallization tendency, which limits the dimensions of the articles and complicates the process of synthesis.

In recent years Belykh had been carrying out systematical investigations with the aim to develop weakly crystallized glasses for crown-flint pairs with high refractive indices and extreme dispersion values. Following the main aim of the present work, it was necessary to choose the least crystallizing initial compositions with sufficiently high refractive indexes. According to the rule of crystallizability, for low-crystallizing glasses the decrease of production, the sections of the system should be used which correspond or border on the lines of joint crystallization of two phases.

To produce commercial glasses with high refractive index, two eutectic compositions of the system As–Ge–Se:$As_{22} Ge_{30} Se_{48}$ (I), $As_{13} Ge_{32} Se_{55}$ (II) were taken and the influence of additions of heavy elements from the IV group and tellurium upon their properties, in particular, crystallizability, was investigated. Metal-containing vitreous melts have, as a rule, high crystallization tendency and they badly withstand a repeat thermal treatment. However, the replacement of only 1% of germanium with tin in the region with a lack of selenium causes the increase of refractive index $\Delta n_\lambda = 0.0230$. Lead enhances the refraction index even more: $\Delta n_\lambda = 0.0370$. That's why it was of interest to determine the optimal number of these elements, at which the glasses being produced could remain technological. The compositions and the properties of the investigated melts are given in the Table 2 in the Appendix. Vitreousness and homogeneity of the specimens produced were evaluated by their external appearance (conchoidal fracture) and the type of transmission spectra.

Visually the melts were vitreous based on composition I with the content of tin, to 12 at.%; of lead, to 10 at.%; and on composition II for the addition of both tin and lead to 14 at.%. However, in these melts with tin content exceeding 5 to 6 at.% and lead content from 3 at.%, the increase of the incline of transmission short-wave limits is observed. This is evident of their heterogeneity, for only the glasses are homogeneous in which shift of short-wave limit caused by the introduction of heavy elements goes parallel to the transmission limit in the initial glasses (Figure 45). This is confirmed by investigations of glass crystallizability by the polythermal method. With the insertion of even a small amount of metals into the initial non-crystallizing glasses, the crystallizability is greatly enhanced.

The glasses based on composition II crystallize, however, upon the addition of 1 to 3 at.% Sn or Pb in the more narrow temperature interval and they can be synthesized in sufficient quantities. Even the insertion of two metals simultaneously into this glass composition in the amount equal to 4 at.% (2.0 at.% of each Sn and Pb) — although it reduces crystallization — doesn't considerably increase the refractive index: n of the initial glass = 2.6. Thus the glasses produced won't differ from the existing ones.

The replacement of 1 at.% of selenium with tellurium in the glasses of the As–Ge–Se system leads to the refractive index increase of $\Delta n_\lambda = 0.0126$. To determine the limiting capacities of tellurium insertions into the eutectic compositions of the system As–Ge–Se, the glasses $As_{22}Ge_{30}Se_{48-x}Te_x$ (I) and $As_{13}Ge_{32}Se_{55-x}Te_x$ (II)

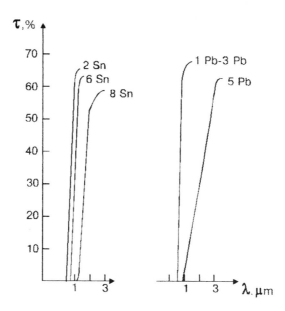

FIGURE 45.   The location of the transmission edge of the glasses $As_{13}Ge_{32-x}(SnPb)_xSe_{55}$.

were synthesized. Compositions and properties of these glasses are given in Table 10. It was established that stable glasses can be produced at tellurium content up to 20 at.%. The fact attracts our attention that tellurium equally influences crystalliz-ability of the glasses of both series.

The further increase of tellurium content causes crystallization growth. In glasses containing 20 at.% this is insignificant surface crystallization, but at tellurium growth up to 25 at.% strong surface and demersal crystallization is observed. At a Te content of 30 at.% the glass crystallizes completely. Glass density increases with tellurium addition, but atomic concentration in a unit of volume decreases, which could be expected due to the significant difference of the covalent radii of Se and Te atoms (1.17 and 1.37 Å, respectively).

As a consequence of tellurium insertion, a shift takes place of both near- and far-transparency edges toward the greater wavelengths and these glasses transmit radiation up to $\lambda = 17.5$ μm. Extrinsic absorption between 12.5 and 14 mm is elim-inated by the extra purification of blend materials (Figure 46, glass 25). Short-wave transmission edge in the glasses with tellurium content from 2.5 up to 20 at.% shifts approximately equally for the specimens of both series: from 0.85 to 0.87, to 1.16 to 1.18 μm; therewith this shift goes without decrease of steepness edge. More gently sloping transmission edge is typical of the melts with tellurium content = 25 at.%, which is evident in the appearance of the second-phase particles in the glass melt (Figure 46, glass 273).

The examination of the lightest glasses under the IR microscope revealed their monophase character. The decrease of transmission value to 60% takes place as a result of the refractive index enhancement.

TABLE 10

Properties of Glasses with Tellurium Additions into the Compositions of the Eutectics of the System As–Ge–Se

Composition (at.%)				T_{ds}	ρ		λ_e	Crystallization after
As	Ge	Se	Te	(°C)	(g/cm³)	$N \cdot 10^3$	(μm)	6 h exposure
22	30	48	—	375	4.492	58.99	—	Does not crystallize
22	30	45.5	2.5	358	4.533	58.55	0.87	Does not crystallize
22	30	43	5	346	4.591	58.4	0.93	Does not crystallize
22	30	38	10	328	4.676	57.77	1.01	Does not crystallize
22	30	33	15	316	4.774	57.76	—	Does not crystallize
22	30	30	18	311	4.833	56.97	1.16	Insignificant demersal crystallization (320–420°C)
22	30	28	20	308	4.882	56.82	—	
22	30	23	25	300	4.97	56.3	1.22	Surface crystallization (420–465°C)
22	30	18	30	292	—	—	—	Continuous crystallization (300–560°C)
13	32	55	—	348	4.439	58.1	—	Does not crystallize
13	32	52.5	2.5	346	4.462	57.5	0.85	Does not crystallize
13	32	50	5	342	4.545	57.34	0.88	Does not crystallize
13	32	45	10	329	4.601	56.61	0.98	Does not crystallize
13	32	40	15	316	4.702	56.16	1.03	Insignificant demersal crystallization
13	32	35	20	308	4.78	55.5	—	
13	32	30	25	300	4.87	54.99	—	
13	32	25	30	—	—	—	—	Continuous crystallization

Despite quite a significant decrease of T_{ds} caused by tellurium insertion, the glasses remain heat-resistant enough; the dispersive coefficient increases with the enhancement of tellurium content. This means that tellurium insertion into the eutectic compositions of the system As–Ge–Se in amounts up to 20 at.% allows one to obtain noncrystallizing glasses available for practical application. They slightly differ in their absolute value, however, from commercial ones.

The replacement of arsenic with antimony increases the refractive index by 0.1 on an average. To create stable glasses in antimony systems the eutectic composition $Sb_{19.2}Ge_{21.4}Se_{59.4}$ was taken as a basis. Crystallizability of the eutectic $Sb_{19.2}Ge_{21.4}Se_{59.4}$ was determined by the polythermal method. A glass specimen weighing 70 g had been exposed for 6 h in a gradient furnace at 600 to 100°C gradient. No crystalline phase was seen in the volume of a specimen even with an IR microscope. In the temperature interval 360 to 400°C, insignificant indicators of the surface change were observed. To determine their resistance to repeat thermal treatment, glass specimens were put through the more continuous annealing. They were at least crystallized out only in a crushed state after 300 h of testing. The

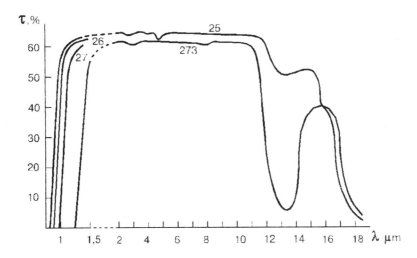

FIGURE 46. Transmission spectra of glasses $As_{13}Ge_{32}Se_{55-x}Te_x$ of different purity.

crystalline phase separated out in the process was identified as antimony triselenide with the help of RPhA. Additional phases were not discovered. Korepanova and others obtained the same result.[166]

To increase the refractive index, additions of heavy elements were introduced into the calculated composition of eutectics. As was noted above, at joint addition of tin and lead into eutectic compositions of the system As–Ge–Se, negative influence of heavy elements on crystallization slightly decreases. To find a correlation of these elements optimal for the system Sb–Ge–Se, the causes of crystalline phase separation on the additions were investigated. Eutectics in the system Sb–Ge–Se is situated in the glass-forming region with the lack of chalcogen, limited by the triangle $GeSe–GeSe_2–Sb_2Se_3$, where glass consists of structural elements $SbSe_{3/2}$, $GeSe_{4/2}$, and $GeSe_{2/2}$. Germanium valency in the last is equal to 2. Then the composition of eutectics is expressed by the following correlation of structural elements: $(SbSe_{3/2})_{19.2}$ $(GeSe_{4/2})_{9.2}$ $(GeSe_{2/2})_{12.2}$. One can expect that only bivalent germanium will be replaced with tin and lead, as this state is the most typical of heavy elements.

As chalcogenides of tin and lead don't have the glass-forming structure, these elements can enter the glass network only when being blocked by the glass-forming structures, i.e., representing not more than a half of the total amount of bivalent germanium. Otherwise elimination of the corresponding selenides is possible — in the form of the crystalline phase.

Roentgenographic analysis of completely crystallized specimens has shown that the products of complex glass crystallization are determined by appearance and concentration of the elements introduced. Thus, when analyzing the products of the initial glass in the composition of which only atoms Sb, Ge, and Se are included, a series of diffraction lines is observed, position and relative intensity of which correspond to the compound Sb_2Se_3, which coincides with the data earlier obtained of crystallization fields of the system Sb–Ge–Se. On the addition of 2.5 at.% Sn into the composition of the eutectic in place of germanium, the compound Sb_2Se_3 manifests itself in crystallization products only on blends' level. Diffraction line with interplanar distance 5.93 Å indicates the presence of the crystal $GeSe_2$. Besides, the phases GeSe and SnSe of orthorhombic syngony are identified. For the compounds GeSe and SnSe, however, changes are observed in the values of interplanar distances and correlation of some lines' intensity in comparison with those which are presented in the file LCPDS. This is the consequence of the mutual influence of both phases' presence as a result of germanium isostructural replacement with tin and vice versa in these compounds. Abrikosov and others suggest in these works that such replacement is possible.[167,168]

On 5 at.% Te introduction into the glass in place of selenium, a considerable amount of $GeSe_2$ is noted in crystallization products. Besides, lines appear corresponding to the cubic modification of the compounds GeSe and GeTe. The absence of lines typical of the compound Sb_2Se_3 is evident of the fact that tellurium, as well as tin, decreases antimony crystallization. True, the glass, containing 2.5 at.% Te, doesn't crystallize at adopted cooling regimes.

In the glass containing 2.5 at.% Pb, inserted through germanium replacement, the lines may be set off on the roentgenograms typical of Sb_2Se_3, $GeSe_2$, GeSe, and PbSe. In this way Pb "demolishes" the glass, preserving antimony crystallization.

At the joint introduction of Sn and Pb — 2.5 at.% each — in place of 5 at.% Ge and 3 at.% Te in place of Se, the principal crystalline phases were orthorhombical GeTe and Sb_2Se_3. As the content of all these elements increases, one can discover the lines appropriate to the compounds PbSe, SnSe, SnTe, Sb_2Se_3, and others. These conclusions are in accordance with the investigations of crystallizability of the same glasses. The results of these investigations are given in Table 11. As expected, the glasses containing, simultaneously, tin, lead, and tellurium, appeared to be the least crystallizing. Their crystallization interval after 6 h of in a gradient furnace doesn't exceed 150°C and lies in the range of temperatures which allows synthesis in sufficient amounts. Thus, in the result of the performed investigation, the glasses of four compositions appeared to be of the best quality technologically. As their properties are not standardized and not marked in technical documentation, they are considered as experimental ones and their index is AV.

The glass AV1 is the unique one among commercial ChG due to the combination of refractive index and dispersion ($n_{10.0} = 2.9920$; $v_{10.0} = 120$), although it lies in proximity of IRG32. This is an IR superheavy flint. The glass AV2 should be considered the crown one. Its refractive index differs only slightly from BS2, but its dispersion is higher ($n_{10.0} = 2.8424$; $v_{10.0} = 162$). The glass AV3 may be compared to heavy flints ($n_{10.0} = 2.7514$; $v_{10.0} = 96$). It has no analogues among commercial glasses. The glass AV4 has no analogues by optical properties not only among commercial, but also among all known ChG. This is IR superheavy crown ($n_{10.0} = 3.0128$; $v_{10.0} = 279$).

It is known that the application of superheavy crowns in optical devices enables one to correct much greater aberrations than at the same number of lens. We should note once more that all glasses of the AV series don't crystallize at melting and repeated heat treatment. Their location on the Abbe diagram $n_{10.0}$-$v_{10.0}$ and physicochemical properties are shown in Chapter 4.

TABLE 11

Crystallization of Glasses of Eutectic Composition of the System Sb–Ge–Se with Additions of Te, Sn, Pb

Composition of glass (at.%)						T_{ds} (°C)	Temperature interval°C of crystallization (hold time 6 h exposure)	RPhA data	Annotation
Sb	Ge	Se	Te	Sn	Pb				
19,2	21.4	59.4	—	—	—	262	330–430	$GeSe_{2ort.}$, GeTe, GeSe	
19,2	21.4	54.4	5	—	—	—		$GeSe_2$ (SnSe, GeSe)	Glasses are obtained at quenching from 550°C
19,2	18.9	59.4	—	2.5	—	232		Sb_2Se_3($GeSe_2$, GeSe, PbSe)	
19,2	18.9	59.4	—	—	2.5	247			
19,2	20.4	59.4	—	0.5	0.5	245			
19,2	19.4	59.4	—	1	1	242			
19,2	16.4	59.4	—	2.5	2.5	232			
19,2	11.4	54.4	5	5	5	—		PbSe, SnSe, Sb_2Se_3	Glasses are obtained at quenching from 550°C
19,2	11.4	56.9	2.5	5	5	—			
19,2	16.4	56.4	3	2.5	2.5	—		$GeTe_{ort}$, Sb_2Se_3	
19,2	17.4	54.4	5	2.5	1.5	—	280–330		
19,2	16.4	56.4	3	3	2	—	296–430		
19,2	16.9	56.9	2.5	2.5	2	—	320–410		
19,2	16.9	57.3	2.1	3	1.5	—	310–436		

3.6 OPTICAL GLASSES FOR PASSIVE ELEMENTS OF CO_2 LASERS[25-27]

Development of new laser systems depends to a large extent on availability of materials which can, for a long time, stand the exposure to high-powered radiations without damage or changing optical properties. That's why much attention is paid to the study of the processes of laser radiation interface with various materials, mainly with optical crystals. Causes and character of damage and the dependence of threshold radiation power on the substance state are under study; damage mechanisms continue to be proposed.

By threshold power we mean the minimum energy value of radiation at which the material decomposes. The morphology of damage of different substances varies therewith and depends not only on the substance itself, but also on radiation parameters. At continuous radiation thermal demolition (burning, cracking) of a specimen volume is, as a rule, observed. As was established in the works by Darvoid and others, who had been studying the influence of radiations of CO_2 laser on the crystals KRS-5 and KRS-6, the attack of impulse radiation causes damage of, first of all, the surface of the material.[169] In the majority of work damage is detected visually: by the appearance of a flare or, less commonly, by transparency loss, fixed by oscillograph. In consequence of an investigation of optical strength of a series of crystalline IR materials performed by Horrigan and others, and also in Deutsch's review, it was demonstrated that the main characteristics of optical materials for laser are transparency in the desired spectrum region and absorption coefficient.[170,171]

The possibility of ChG applications to CO_2 lasers is noted in the work by Sparks,[172] and in publications by Hilton,[173] Lezal,[154] and Moynihan.[153] Not one of these works, however, presents either the data for the character of interaction of laser radiation with glass or the values of damage limits. Systematical investigations of the dependence of absorption coefficient and limiting powers for ChG damage on the composition and quality of glass and also development of glasses for passive elements of IR lasers, described in this part, have been carried out by Kislitskaya in the beginning of the 1970s.

3.6.1 OPTICAL ABSORPTION AND RAY STRENGTH OF THE GLASSES As–Se, Ge–Se, AND SELENIUM

Initial investigations of ray characteristics were carried out as applied to vitreous selenium and glasses of the binary systems As–Se and Ge–Se. Vitreous selenium and glasses of the system Ge–Se were synthesized in the laboratory — 500 g of each composition simultaneously. The glasses of the system As–Se were produced in an amount not less than 1 kg. The specimens were made of glasses, taken from all the meltings, for quality control under the microscope and spectrophotometrical measurements and also for the investigation of absorption coefficient and radiant strength. The specimens were polished flat-parallel plates $44 \times 40 \times 8$ mm^3 or disks Ø50 mm, 8 ± 0.2 mm thick. Under the IR microscope at 280X magnification dark insertions were disclosed, and their dimensions varied in glasses of different systems. Thus, crystalline insertions in selenium had the form of spheroids. On the contrary, in glasses of the As–Se system there were no clearly defined crystalline formations, but there were many of evenly spread small insertions (2 to 6 μm). In glasses of the Ge–Se system very large (20 μm and more) unevenly spread particles of vague form were observed. As our experience shows, the insertions similar to those observed in glasses of the As–Se system present, as a rule, the result of crystallization of the glass components or their oxygen compounds. True, we managed to eliminate most of these insertions, by raising the quenching temperature above the liquidus one and purifying the initial materials from oxygen admixtures more accurately.

In glasses of the Ge–Se system the first crystalline phase represents well-identified short-prism crystals $GeSe_2$. We have supposed that the observed defects were the result of incomplete melting and we were right: the glasses with a low number of insertions were obtained after the duration of melting had been increased and due to the more intensive mixing of the melt.

Composition of heterophase insertions couldn't be conclusively established, for there were no methods of nondestructive control. The method put forward which allowed us to investigate the dispersion composition and concentration of insertions in the glasses As–S and As–Se was laser ultramicroscopy using a He–Ne laser. To investigate selenium glasses, which are opaque at this wavelength, dissolving a specimen was required. That is why the nature of the second phase in ChG, hence its origin, can be defined only positively.

To find out how the defects revealed influence spectral characteristics of absorption coefficient, these values were specially measured for contaminated specimens. The results of microscopic and spectral investigations are given in Table 12. One can see that transmissions in the short-wave part of the spectrum correlate with the amount of dark insertions. Position and intensity of absorption bands in wavelength interval 10 to 15 μm are determined mainly by the composition of glasses. Thus, in the glasses $As_{10}Se_{90}$, $As_{20}Se_{80}$, and $Ge_{10}Se_{90}$ the absorption band about $\lambda = 13.6$ μm is observed, stemming from intrinsic vibrations of bonds Se–Se. As the content of arsenic increases up to 30 at.%, transmission decrease takes place from $\lambda = 13.6$ μm, stemming from imposition of selenium absorption band and the bonds As–Se (14.3 μm). The glass $Ge_{20}Se_{80}$ with a large amount of dark insertions has not only lowered transmission all over the entire spectrum, but also a large absorption band at 12.8 μm, which is identical to oscillations of the bond Ge–O. This suggests that in the given specimen there was also the greatest amount of admixtures in the form of germanium oxides.

At the wavelength 10.6 μm all the glasses besides $Ge_{20}Se_{80}$ have a maximum transmission.

The absorption coefficient is one of the main factors which influence the laser damage limit. It is identified as the relation of the amount of radiation, absorbed by a solid, to the amount of radiation which was incident on it. In the IR region only high values of absorption coefficients can be identified by the spectral method and with low accuracy at that, owing to the error introduced by distortion of a bundle of light by specimens over 10 mm thick.

As a rule, the spectral method is applied to establish the absorption coefficient, i.e., the reciprocal of the distance at which the current of monochromatic radiation

TABLE 12
The Quality of the Glasses As–Se, Ge–Se, and Selenium Influence upon the Absorption Coefficient and Threshold Radiation Power

| Glass composition (at.%) | Dark insertions | | Transmission at λ (μm) | | | $K_{10.6}$ cm^{-1} | $\mathfrak{S}_{thr}$ (W/cm^2) |
	No. pieces/cm^3	Dimensions (μm)	10.6	12.8	13.6		
Se	3,000	10–30	68	46	32	0.06	—
$As_{10}Se_{90}$	~9,000	10–40	68	64	43	0.028	—
$As_{20}Se_{80}$	7,500	10–50	68	60	46	0.028	9.7
$As_{30}Se_{70}$	3,700	10–40	68	60	47	0.025	—
$Ge_{10}Se_{90}$	2,600	30–150	69	60	41	0.018	10
$Ge_{20}Se_{80}$	>10,000	10–100	64	32	50	0.028	6

is reduced by a factor of ten by the substance. The thicker a specimen is, the higher the accuracy of the measurements.

The measurement of low values of absorption coefficients in the region 10 μm is performed by the laser calorimetric method based on determination of the temperature change of the specimen caused by the absorption of a part of incident radiation. This method, put forward in the works by Horrigan[170] and Artyushenko and others,[174] was used at the investigation of IR crystalline materials.

In the present work all the measurements of the absorption coefficient were performed by the method of laser calorimetry. As the source of radiation we used the quasi-continuous CO_2 laser with a power of about 100 W, and normal laser LG-22 with a power of about 30 W. The measurements were carried out on the flat-parallel specimens $44 \times 40 \times 8$ mm^3, or disks 30 and 50 mm across, 5 and 8 mm thick. A specimen with a junction of Chromel-Alumel thermocouple, fastened to a side, was put into the thermostat. The power of radiation having passed through the specimen was fixed by power gage. The temperature of the specimen was measured with an accuracy of 0.1°C. The absorption coefficient — taking into account heat exchange with the environment and multiple reflection, the account of which must be taken because of the high refractive index — is calculated from the formula:

$$K_{10.6} = \left(c \times m \times \Delta T/L \times P\tau \right)\left(1/t_1 + 1/t_2\right)\ 2n_{10.6}/n_{10.6}^2 + 1$$

where c is specific heat; m is mass of a specimen; $\Delta T = T_2 - T_1$ is a specimen's temperature change in the process of measurement; t_1 and t_2 are duration of the period of a specimen's warming up from T_1 to T_2 and cooling from T_2 to T_1; $P\tau$ is power, having passed through a specimen; and L is the specimen thickness.

The accuracy of the measurements is equal to ~25%. Apart from the method of spectral absorption the calorimetric method can result in slightly overstated values of the absorption coefficients due to admixtures which, as a rule, exist in all real materials. The values of absorption coefficient are given in Table 12. As is seen from the table, the absorption coefficient is quite high and, unlike spectral characteristics, it doesn't depend on glass composition.

Threshold destruction power (I_{thr}) in the continuous regime was measured on the same lasers as K_{abs} at the diameter of the spot being irradiated from 1.5 to 24 mm. Reduction of radiation intensity was carried out with the help of a set of absorption filters, focusing of radiation into spots of small diameter — with the help of a NaCl lens. Although it was obvious that at high values of $K_{10.6}$, obtained for selenium and glasses of the systems As–Se and Ge–Se, threshold destruction powers would be low, it was appropriate to investigate character and nature of laser destruction on these very specimens.

Specimens in the form of disks Ø50 mm, 8 ± 0.2 mm thick were placed into a laser beam ray with different power density. Destruction was observed visually. In the process of testing it was established that at introduction of a specimen into a laser beam Ø10 mm, a part of it which is exposed to radiation attack at once warms up, whereas the edges remain at room temperature due to the low heat conductivity of ChG. As a consequence of the formation of a temperature gradient cracks appear and destruction of a sample occurs — which is apparently of a thermal nature — and apart from the low heat conductivity it is aided by high coefficients of thermal expansion, typical of ChG. In these conditions threshold destruction powers for glasses of both systems do not exceed 10 W/cm² (Table 12).

When the samples are inserted in the path of a beam 1.5 to 2 mm across, the character of selenium and binary glasses destruction is different. A selenium sample had been withstanding power density 300 W/cm² for 5 min without destruction. At power density increase up to 500 W/cm² fusion of radiated zone was observed.

Samples of the glasses As–Se and Ge–Se at the same power density were covered with small cracks and at power increase up to 800 W/cm² the glass inflamed, but the samples, as a rule, didn't crack. It seems likely that at this very power quantity breakage of the bonds occurs with substance evaporation prior to temperature gradient formation. We can suppose, however, that the breakage of bonds is aided by prompt warming up of the irradiated volume, which takes place due to the energy absorption on insertions of different kinds. The analysis of transmission spectra and the results of microscopic study of the samples demonstrates that insertions in the glass volume appear in consequence of incomplete melting of the charge components or entrapment of foreign pollutions. Among extrinsic absorption bands observed in ChG, the bands, caused by oxygen combined in glass, in the region 9.6 and from 12.5 to 13.3 µm may be considered to be the closest to the frequency of generation (943 cm⁻¹). Besides, in the spectrum of arsenic selenide the existence was noted of one more weak band in the region 10.6 µm. Moynihan explains it by the presence of bonds Se–O.[153] In the samples which we investigated, the extrinsic band about 12.8 µm was observed only in the glass with 20 at.% of germanium. Other bands are either absent or not resolved due to their low intensity. That's apparently why high values of the absorption coefficient are mainly caused by absorption on insertions. Extrinsic absorption bands, however, influence to some extent the absorption coefficient, in this case in the glasses Ge–Se the band 12.8 µm numerous bands from 8.6 to 14 µm, due to oxygen bonds with arsenic.

In this way, low values of threshold destruction power were established for vitreous Se and binary glasses As–Se and Ge–Se, which have a considerable amount of different pollutions. The value of threshold power in continuous working regime of the generator at great dimensions of a radiation spot doesn't depend on glass composition.

The measurements of threshold destruction capacity were performed in the pulse regime on CO_2 laser with pulse duration 10^{-7} s at the area of irradiated spot 1×10^{-3} cm^2 on the flat-parallel polished samples. The values of threshold capacity were calculated as arithmetic mean from five to ten measurements on each sample. The investigations were carried out on those very examples which were used in continuous regime. Prior to testing the surfaces were rubbed first by a damp napkin and then by a dry one. Damage threshold attained was fixed visually when spark plasma flame appeared, which caused a breakthrough in the surface.

For pulsed radiation the threshold of destruction of a processed surface is always lower than that of its volume. Views of a surface destruction copied at examination under a microscope in reflected light at various magnifications are presented in Figure 47. One can see that on the surface of the glasses $As_{20}Se_{80}$ and $Ge_{20}Se_{80}$, for the most part these are puncturings (Figure 47a), but at values very close to the threshold ones, destruction is observed in the form of congregation of a spherical crater (Figure 47b).

On the surfaces of glasses $As_{30}Se_{70}$ and $Ge_{10}Se_{90}$ sequential stages of destruction are traced with the increase of radiation intensity: at first this is the congregation of small spherical craters, then cracks appear around them. Above damage threshold, puncturing of glass occurs. At examination of samples in transmitted light no changes in volume were stated.

Further checking was performed of how the quality of a material cool treatment influences the quantity of glasses' surface rigidity. For that one plane of a sample was processed according to the standard technique, which ends with polishing on the tar polisher by water suspense Cr_2O_3 (normal grinding and polishing, NGP). The other plane of a specimen was processed in the regime of deep grinding and polishing (DGP), when the whole cracked layer is removed and then polished on the tar polisher by diamond powder $M_{0.5}$. Samples of selenium, due to its elasticity, were polished on the velvet polisher by water suspense Cr_2O_3 with diamond paste. To avoid the error caused by the possible formation of oxide coating on the surface, the polishing was carried out just before the measurements.

Testings of surfaces, processed in different regimes, demonstrated that the character of destruction doesn't depend on the degree of surface finish; however, with its refinement the beam intensity increases up to 40% (Table 13). The data given in the table were obtained as average ones from five to ten rigidity measurements on one surface. This phenomenon leads to the necessity to check the ChG ability to undergo "laser purification", established by Kovalev and others for crystals,[175] which resides in rigidity enhancement as the radiation of constantly increasing power attacks one and the same point of the sample. The authors note that in consequence of such purification the breakthrough threshold of the surface of crystal may increase up to the threshold of volume destruction. It is supposed that in the process of laser purification residues of easily burning organic pollutions are easily removed from

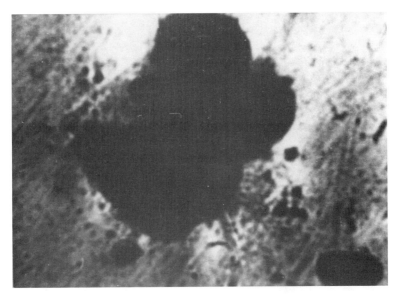

FIGURE 47a(I).

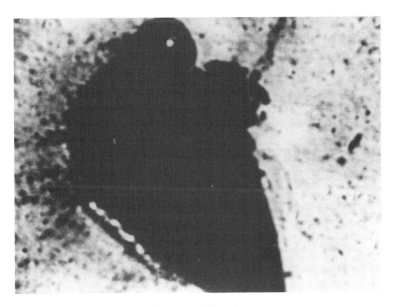

FIGURE 47a(II).

FIGURE 47a,b(I) and (II).   Decomposition of glass surfaces by action of pulseed radiation of CO_2 laser for 0.1 μs. (I) Evaporation craters (a) and puncturing (b) on the glass $As_{20}Se_{80}$; (II) evaporation craters (a) and puncturing (b) on the glass $Ge_{20}Se_{80}$.

the surface, sorbed moisture evaporates, and the roughness even gets smoothed to some extent. The experiment showed that in this regard glasses are also identical to crystals (Table 14). One can see that as a result of laser purification the destruction

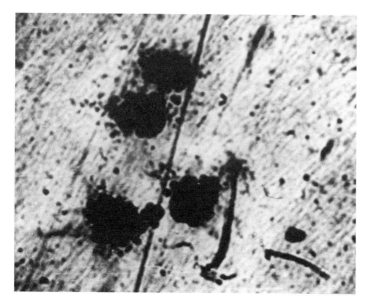

FIGURE 47b(I).

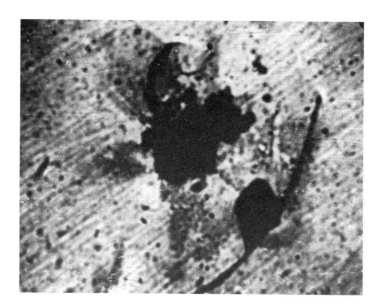

FIGURE 47b(II).

FIGURE 47a,b(I) and (II) Continued.

power increases approximately twice as much. Therewith the worse the first mechanical treatment is, the more distinct is the effect of laser purification, which confirms the supposition about smoothing of the surface roughnesses.

TABLE 13
Dependence of Beam Strength on the Degree of Glass Surface Finish

Classes composition (at.%)	Type treatment	Surface strength (MW/cm²)	$K_{10.6}$ cm^{-1}
$As_{20}Se_{80}$	NGP	41 ± 11	} 0.028
	DGP	57 ± 12	
$Ge_{20}Se_{80}$	NGP	37 ± 8	} 0.028
	DGP	52 ± 5	

TABLE 14
"Laser Cleaning" Influence on Beam Strength

Classes composition (at.%)	Surface finishing	Surface beam strength (MW/cm²); subsequent increase of power radiation	Decomposition power at "laser cleaning"
Se	On velvet	$47 \rightarrow 52 \rightarrow 72 \rightarrow 81 \rightarrow 90$	99
$As_{30}Se_{70}$	NGP	$37 \rightarrow 45 \rightarrow 53 \rightarrow 56 \rightarrow 66$	68
$Ge_{20}Se_{80}$	DGP	$24 \rightarrow 27 \rightarrow 28 \rightarrow 38 \rightarrow 50$	59

ChG are nonhydroscopic, but when polishing by water suspension a monomolecular layer of moisture may remain on the surface. In this connection inspection was carried out of the threshold power dependence on composition of the liquid used for surface purification.

The testings were carried out at sequential treatment of one and the same surface by water, 100% ethanol, and carbon tetrachloride. The data given in Table 15 demonstrate that ethanol washing decreases radial rigidity even in comparison with water, whereas the treatment with CCl_4 leads to its enhancement. Apparently at radiation combustion occurs with heat liberation of remaining ethanol traces on the surface. Meanwhile evaporation of moisture remainders requires energy absorption. The treatment by CCl_4 removes traces of moisture and organic pollutions by itself and that leads to enhancement of threshold power of decomposition.

As was demonstrated above, in the glasses investigated a sufficiently large number of dark insertions were observed, determining high values of absorption coefficient. It could be expected that these insertions caused decomposition of glass surfaces at the pulsed laser attack. However, as is seen from Table 16, no correlation exists between $K_{10.6}$ and surface radial strength. At the same time the highest values of radial strength in selenium, a clear tendency to its reduction in germanium glasses if compared to the arsenic ones, and also the change of decomposition character

TABLE 15

**The Influence of Washing Liquid
Composition on Beam Strength**

Glass composition (at.%)	Surface finishing type	Washing liquid	Surface beam strength (MW/cm²)
Se	On velvet	Water	58 ± 8
		Alcohol	45 ± 2
As$_{10}$Se$_{90}$	NGP	Water	52 ± 8
		Alcohol	41 ± 3
		CCl$_4$	60 ± 9
Ge$_{20}$Se$_{80}$	DGP	Water	47 ± 12
		Alcohol	35 ± 8
		CCl$_4$	55 ± 11

with the composition change are evident of the glass structure influence on the quantity of the decomposition threshold. It can be supposed that under the influence of short pulse in vitreous selenium, the deformation of chains without bonds' breakage initially occurs. The system seems to exhibit resistance to the action of the impulse. As the power increases, a prompt warming up occurs of a very small area near the surface layer, and decomposition in the form of puncturings, similar to the action of continuous radiation, appears in consequence of the temperature gradient formation. Due to that the threshold power of decompositions appears to be higher than that appropriate to the energy of the breakage of the bonds Se–Se.

TABLE 16

**Correlation between Surface Beam
Strength and Absorption Coefficient**

Glass composition (at.%)	Beam strength (MW/cm²)	Absorption coefficient ($K_{10.6}$cm^{-1})
Se	58 ± 8	0.06
As$_{20}$Se$_{80}$	41 ± 11	0.028
As$_{30}$Se$_{70}$	51 ± 16	0.025
Ge$_{10}$Se$_{90}$	32 ± 4	0.018
Ge$_{20}$Se$_{80}$	37 ± 8	0.028

On arsenic introduction into the glass network there appear trigonal elements AsSe$_{3/2}$, less elastic than selenium chains. Spatial tetrahedral structural elements GeSe$_{4/2}$ produce even more rigid structure. The experiment shows that as elasticity of the system reduces, the quantity of radiation power, necessary for the breakage of bonds, decreases. In these cases, when the threshold power is achieved, the evaporation of glass occurs with formation of craters having the form of a dissected bubble. At the further enhancement of power mechanical destruction begins in the form of cracks and puncturings, also of thermal nature. Thus, despite the higher energy of the bonds As–Se (52 kcal/mol) and Ge–Se (56 kcal/mol), if compared to the bond Se–Se (49 kcal/mol), smaller destructive energy is needed for their decomposition (at the expense of elasticity reduction).

3.6.2 OPTICAL ABSORPTION IN GLASSES OF COMPLEX COMPOSITIONS

The next stage in the solution of the problem of using ChG as passive elements of laser systems was the study of radial strength of complex glasses. To get rid of the influence of various admixtures on $K_{10.6}$, naturally needed are the careful purification of raw materials and digesting vessels, reduction to a minimum of 1 ppm of oxygen content in digesting atmosphere, etc. It is known, however, that the presence of oxygen is already enough for the absorption bands stemming from the bond Me–O to appear. Besides, the absorption on the bonds Me–Se and Se–Se can also influence the absorption coefficient. Hence it is desirable to shift the absorption bonds — both extrinsic and intrinsic — toward the greater wavelengths, moving them as far as possible from the generation frequency of the CO_2 laser.

The shift of absorption bonds is achieved by the introduction of heavier elements into the glass composition. The arsenic-free glass IRG29 is the most interesting from this point of view. The measurements of the absorption coefficient of the glass IRG29 were carried out on the specimens of commercial meltings, made of ignots weighing 4.5 to 5 kg, synthesized in usual regime. Characteristics of the glasses are given in Table 17.

On the specimens of glass of the founding No. 3705 the influence of a specimen size on the quantity $K_{10.6}$ has been studied, and the error of measurements was assessed of the absorption coefficient in ChG by the laser calorimetric method. For this purpose four series of samples were prepared — five to six pieces in each, 30 and 50 mm across, 5 and 10 mm thick. All the samples were prepared simultaneously in working conditions by the method of double molling and passed through thin annealing with a cooling speed of 2.5 to 5°C/h. That's why residual pressures were the same in all the specimens. In all cases the diameter of the laser beam was 24 to 30 mm.

As far as any difference in technological past history is excluded, the observed scatter of values $K_{10.6}$ is apparently characteristic of the method's accuracy. The average error in the measurements given in the reference literature, calculated from the formulas, was 18% and it concurs with the data known for the calorimetric method.

Also noted is the dependence of the value $K_{10.6}$ on the size of a sample: on the samples with diameter 30 mm, when the sizes of a sample and a radiation beam are comparable, the average values of the absorption coefficient are higher than on the samples of large diameter. This may be explained by a partial ingress of scattered radiation on a thermocouple, placed on the side surface of a sample. $K_{10.6}$ of the

TABLE 17
The Dependence of Beam Strength of IRG29 of Commercial Meltings on Glass Quality

| NN meltings | Weight (kg) | Sample thickness (mm) | Transmission ($\tau_\lambda\%$) | | Insertions | | Absorption coefficient ($K_{10.6}cm^{-1}$) | Threshold radiation power ($\mathfrak{I}_{thr}$ W/cm^2) | Time until decomposition (s) |
			$\lambda = 10.6$ μm	$\lambda = 12.7$ μm	Number pieces/cm^3	Size (μm)			
3705	5	8.2	68	58	~5000	10–30	0.018	24	600
3700	4.5	8.8	69	45	~1100	10–40	0.012	30	300

glass No. 3705 was also measured in FIAN. Therewith the change of transmission of CO_2-laser radiation was fixed by one and the same glasses in two specimens 15 and 50 mm across. The obtained value of $K_{10.6}$ was 0.017 cm^{-1}, which coincides with the average one that we had measured on the specimen 50 mm across, 5 mm thick, and equal to 0.018 cm^{-1}. $K_{10.6}$ of the glass No. 3700 is equal to 0.012 cm^{-1}. The difference in absorption quantities is determined by the number of dark insertions, but the absolute value is large enough.

It can be considered that in glasses with an excess of selenium — even without any insertions — the absorption coefficient will hardly be considerably lower than 0.01 cm^{-1}, for, according to the data by Lezal and Srb, the absorption band about 10.5 μm conforms to oscillations of the bond Se–Se. This band, although insignificant in size, is in the region of CO_2-laser radiation and it must be of significant influence on the value of the absorption coefficient, hence of the threshold radiation power.

Glasses exist, free of the bond Se–Se in the system As–Ge–Se in the region of the mixed structures with the bond Ge–As — limited by the content of Ge = 20 to 35 at.% and As = 15 to 35 at.% — and also in the region of ternary eutectics of the system Sb–Ge–Se. Such glasses must have low intrinsic absorption at a frequency of CO_2-laser radiation, because the bands stemming from the oscillations of the bonds As(Sb)–Se. Ge–Se, and As–Ge lie in the long-wave range (from 14 μm). They are characterized by the comparatively low values of CTE ($\approx$12 to 17×10^{-6}/°C^{-1}) and the highest T_{ds} ($\approx$400°C) among the selenium-containing glasses. Taking into account the data for crystallizability, the glass standardized under the marker IRG34 appeared to be optimal for commercial production. Its properties are given in Table 18.

TABLE 18
The Properties of the Glass IRG34

Optical	λ μm	2	9.4	10.6	11.8		$\gamma_{10.6} = (n_{10.6} - 1)/(n_{9.4} - n_{11.8})$				
properties	n_λ	2.631	2.598	2.595	2.591		204				
Thermal			CTE°C^{-1}				T_{ds}°C				
properties			12.5×10^{-6}				375				
Viscosity	lg η	13	12	11	10	9	8	7	6	5	4.5
	T°C	332	360	372	405	428	451	473	498	527	544
Microhardness H(kg/mm²) — 250						Density ρ (g/cm³) — 4.47					

However, the transmission of IRG34, melted in the standard technical process, is worse than that of IRG29: interval 10.2 to 14.1 μm the extremely deep absorption band caused by the position of oscillation frequencies of the bonds As–O and Ge–O is observed.

To take into account easy oxidizability of arsenic and the presence of extrinsic oxygen in selenium, the difficulties connected with this band's removal become apparent. Apparently, for the proper evaluation of the glass adaptability for lasers, it was necessary to work out the technological process of their production in the especially pure state.

3.6.3   THE PECULIARITIES OF SYNTHESIS OF CHG FOR LASERS.
METHODS FOR MANUFACTURING PURE GLASSES

The prime concern for the investigators who pursue the goal of reducing optical losses in ChG — in particular for their possible application in the laser optics — is, as is seen from the above, in removal of extrinsic absorption bands and dark insertions. A series of synthesis techniques is known which allows, in some way or another, to solve this problem. Thus, Savage and Nielsen,[81,145] investigating different ways for oxygen removal — namely, washout and calcination of quartz ampules, washing up of components' blend at evacuation, the synthesis temperature rising up to 1000°C, germanium melting in hydrogen current — came to the conclusion that the best results, if compared with normal vacuum synthesis, are obtained at glass distillation in hydrogen current. Lezal and Srb,[155] to produce the binary glasses As–Se and Ge–Se without oxygen admixture, suggest performing purification of initial components just before melting by elimination of oxides and subsequent distillation of selenium into the melting vessel with germanium or arsenic. The technique applied by Hilton until 1980, was based on the same principle of initial component distillation into the initial vessel with germanium from two other vessels jointed with it.[173,176] Lezal and Srb also put forward an effective method of As_2O_3 removal out of the glass As_2Se_3, based on the reaction of As_2O_3 interface with the products of thermal decomposition of carbamide. At decomposition carbon monoxide and ammonia deoxidize As_2O_3 to arsenic.[177] Moynihan, investigating the absorption of the glass As_2Se_3 with additions of As_2O_3, came to the conclusion that small additions of As_2O_3 (up to 50 ppm) are extracted through diffusion into the walls of the vessels.[153] Besides, As_2O_3 can be moved out of As_2Se_3 by joint melting of the last with B_2O_3. In a series of works techniques are described based on carbon and some metals' ability to absorb oxygen. Among them are the investigations by Brau, who suggested to apply thin carbon coating to the inside surface of the founding vessel,[178] and by Srb and Wachtl, who placed carbon in the path of vapors of the distilled glass.[179] Small additions of aluminum as a getter were proposed by Hilton and Lezal. We reproduced the majority of these techniques and determined that they have certain drawbacks. Thus, only those compositions are adaptable for glass distillation which don't crystallize with repeated treatment, i.e., if in the process of evaporation and condensation decomposition doesn't occur with nonglass-forming phase appearing. The process of the initial components' distillation into the initial vessel is rather difficult for technical execution, especially in working conditions, and it doesn't exclude the possibility of glass impurity with silica at the removal of quartz membranes between the section. At carbamide interface with As_2O_3 the bubbles H_2Se appear in the glass melt and for their removal the subsidiary distillations of glass are recommended. Some getters in one way or another come in touch with melt and in some amounts enter the glass as dark insertions, reducing its transparency along the whole spectrum. That's why, apparently, the above-described methods didn't fall outside the scope of laboratory experiments.

The original method for extra pure multicomponent glasses was proposed by Reitter and others.[180] It differs from Hilton's method[176] in that, due to the special construction of the melting ampul, the distillation of Se, As, and Te into the zone containing extra pure Ge and Sb is carried out without application of partitions made of porous quartz. The authors apply Mg as getter. The glasses don't differ by degree of purification from those obtained in the previous works.

The methods proposed by Lezal and Srb in 1987[154] and by Hilton in 1986[181] are more technological. Their advantage is that all operations are performed out of contact with air, but the substances used must be thoroughly cleaned from oxygen and moisture. The first method allows us to synthesize only As_2Se_3. It is based on the reaction $2AsCl_3 + 3H_2Se \rightarrow As_2Se_3 + 6HCl$. The reaction vessel is fed with a rigidly controlled number of reagents and the vessel is heated up to 700 to 800°C. As_2Se_3 is condensed in a quartz container at $T = 400$°C and then the container is evacuated, annealed, and digestion of a condensate is carried out at $T = 800$°C. The glasses of complex nonstoichiometric compositions are produced by Hilton's method. On the basis of this method the industrial technology was developed and commercial glasses are being produced, hence it is described in detail in Chapter 4.

In Table 19 the values of absorption coefficient are given, obtained as a result of application of various methods for extra pure glass production. One can see from the table the progress in ChG purification degree since the 1960s up to the present. It can also be seen that the absorption coefficient had never been measured until 1975. As_2Se_3 was produced in the most pure state by the spectral characteristics.

Glass's transmittion spectra analysis from literary data has demonstrated that no technical process modification can help to get rid of extrinsic absorption bands without preliminary purification of raw materials.

Quantitative evaluations of content of admixtures in initial materials and glasses are rather rare. In 1989 to 1990 works appeared by Borisevich, Plotnichenko, and others,[149,151,152] where the coefficients of extinction of the groups S–H, Se–H, and Se–O were evaluated. It allowed one to determine the concentration of admixtures through the intensity of absorption bands. There are no data, however, for the groups Ge–O, As–O, and Sn–O which are in charge of the primary band at $\lambda = 12.8$ μm. The work by Ajio and Shkalkova, through neutron activation analysis, determined the total oxygen content in ChG falling within the limits 0.005 to 0.58 mass% and linear dependence of oxygen amount on the intensity of the band 12.5 to 13 μm. Also worth noting are the data obtained by Vikhrov and others,[182] where the total carbon content in pure glasses on the level 0.002 to 0.018 mass% is established. As follows from the review presented by Devyatikh and Churbanov, referring to 1990,[183] the method of neutron-activation analysis, despite its complicated and specific techniques, is the most promising for defining various admixtures in ChG. Besides, to define oxygen, carbon, and hydrogen, gas chromatography is very effective, which allows one to reveal hydrocarbons, carbon oxides, and sulfides in constant 10^{-6} to 10^{-8} mass%. These methods are, however, rather complicated for everyday use and, at the development of the technical process for laser ChG synthesis, we oriented to spectral analysis and laser calorimetric method for $K_{10.6}$ defining.

The advancement of the synthesis techniques, used by us, began with the purification of raw materials. As polycrystalline germanium and metallic tin are obtained through the zone melting method, their supplementary purification isn't necessary. The technical process for obtaining As, Sb, and Se, usually applied at chemical enterprises, doesn't assure the absence of oxygen admixtures in them, and these

TABLE 19
Modification in the Process of Vacuum Synthesis (Review)

Investigated glasses	Peculiarities of technological process	Results of the synthesis	Absorption coefficient (cm⁻¹)	Ref.
As–Se, Ge–Se, As–Ge–Se	Vacuum synthesis	Absorption bands in the region 8–15 μm	Is not measured	81
$Ge_{20}Se_{80}$, As_2Se_3, $Ge_{30}As_{20}Se_{50}$	Distillation in hydrogen flow	Composition change no bands 8–15 μm	Is not measured	145
Glasses with S, Se, Te	Distillation with carbon purification of vapors	No bands in the range 2–25 μm	Is not measured	179
As–Se, Ge–Se	Se distillation into the compounding vessel with Ge or As	No bands at $\lambda = 12.7$ μm	Evaluation 0.01	155
As_2Se_3	The method of carbamide decomposition	Oxygen removed; no bands in the range 2–25 μm	~ 2–$5*10^{-3}$	177
As_2Se_3	Synthesis from gaseous phase	Pure glasses, without bands or insertions	Evaluation ~ 0.001	154
As–Se, Ge–Se, P–Ge–Se, P–Se, As–Ge–Se, and others	Vacuum synthesis	Absorption bands in the range 8–40 μm	Is not measured	80
As_2Se_3	Vacuum synthesis; thorough purification of the ampul, reactants warming up; melting with B_2O_3	Absorption in the range 12.5–15.5 μm decreased three times; no other bands	Laser colorimetry 0.01	153
Sb–Ge–Se	Se and As(Sb) distillation into the vessel with Ge, getter — Al	Band at 12.7 is removed practically completely	$K_{min} = 0.008$	173
Tl1173, Tl20	Se and As(Sb) distillation into the vessel with Ge, getter — Al	Band at 12.7 is removed practically completely	Tl473 K = 0.01; Tl20 K = 0.05	176
Glasses, AMTIR	Synthesis and then glass distillation through the quartz filter without atmosphere contact	No absorption bands except H_2Se in the range 4–5 μm	$K_{4-6} < 1 \times 10^{-3}$; $K_{10.6} \sim 0.007$	181
Ge–As–Sb–Se–Te	As, Te, Se distillation into the vessel with Ge, Sb, not using quarts filter, getter — Mg	Doesn't differ by purity from the glasses (Hilton[173])	Is not measured	180

chemicals were supplementary purified prior to the synthesis of glasses. As a result the glasses IRG29 and IRG34 were synthesized with unit insertions 2 to 4 μm in size, evenly spread in the volume. The investigations performed on the electron microsound installation "Sameca" have demonstrated that the insertions contain silicon — possibly as SiO_2 — which, according to different authors' data, can be connected with glass from the walls of the founding vessel in the process of synthesis.

The influence of the absorption bands in the frequency region 8 to 14 µm on the absorption coefficient was investigated on samples 7 to 8 mm thick, prepared from such glasses. Chosen were the glass IRG34 of five foundings with different transmission in the absorption band λ–12.8 µm and the glass IRG29, where there was practically no absorption band. The results are given in Table 20 and in Figure 48.

TABLE 20

The Influence of Extrinsic Bands in the Range 8–14 µm on the Absorption Coefficient $K_{10.6}$

Glass type	Sample no.	Result of microscopy	Transmission ($\tau\lambda\%$) $\lambda = 10.6$ µm	$\lambda = 12.8$ µm	$K_{10.6}$ cm^{-1}
IRG34	1	Insertions are not viewed	67	4	0.017
	4	Unitary small insertions	67	13	0.006
	5	Many insertions of various sizes	59	15	0.085
	2 ⎫	Unitary insertions up to	67	20	0.008
	3 ⎭	4 µm in size		53	0.005
IRG29	6	Small insertions exceeding in number those in samples 2–4	69	60	0.014

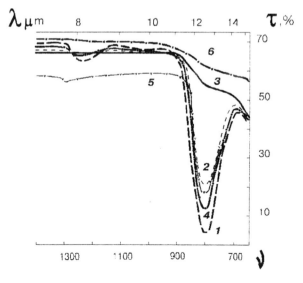

FIGURE 48. Transmission in the range 8 to 15 µm of the glass IRG34 of different meltings and of the glass IRG29. (1–5 — IRG34; 6 — IRG29.)

The experiment has demonstrated that the absorption band with the center at 12.8 μm influences $K_{10.6}$ only if the transmission in the center is equal to zero (sample 1). The quadruple difference in transmission of samples 3 and 4 almost doesn't influence the value $K_{10.6}$. However, the sample of the glass IRG29 which slightly differs from sample 3 by the number of insertions and transmission in the absorption band 12.8 μm has the absorption coefficient three times as much. It can be explained only by the presence of the excessed selenium in the composition of glass; and, nevertheless, the number of insertions predominantly influences the absorption coefficient. It can be seen that sample 5, taken for comparison out of a deficient founding, has the same transmission in the absorption band as sample 4, but differs by the great number of insertions. As the presence of a number of second-phase particles of any character reduces the transmission of glasses along the whole spectrum, it's quite natural for $K_{10.6}$ to be rather considerable in this sample (0.085 cm^{-1}). Analysis of the results of the investigations of all the glasses shows that the value $K_{10.6}$ doesn't depend on composition. The same can't be said about the technological peculiarities of synthesis, and it also determines the quality of glass.

Apart from the purification of raw materials, the time-temperature melting regime was thoroughly corrected, and these techniques together allow us to produce the glass IRG34 in working conditions with stable value $K_{10.6} = 5$ to 6×10^{-3} cm^{-1} without further devices (Table 21).

TABLE 21
The Dependence of Beam Intensity on the Quality of Glass IRG34 from Commercial Meltings

Melting no.		Thickness of samples (mm)	Transmission $\tau_\lambda\%$ ($\lambda = 12.8$ μm)	No. of insertions piece/cm^3	Absorption coefficient ($K_{10.6}$ cm^{-1})	$\Im_{thr}$ (W/cm^2)	Time to decomposition (min)
1156	a	9.46	12	—	0.005	40	Defocused
5459		8	22	70	0.005	40	16
5457		8	33	100	0.006	40	Defocused
1156	b	7.45	16	350	0.005	30	15
3416		7.8	53	500	0.005	30	10
1156	c	7.15	19	500	0.008	30	11
3565		7.8	20	500	0.008	25	10
4853		7.8	13	700	0.007, 0.008 (FIAN)	27	10
4911		7.8	37	1600	0.01	30	15
3565		7.85	20	>10.000	0.085	13	1 s

3.6.4   RADIAL STRENGTH OF THE GLASSES IRG29 AND IRG34

The dependence of radiation thresholds on glass quality was established on the same samples with the absorption coefficient. The measurements were performed in the homogeneous laser bunch 10 to 13 mm across, using a CO_2 laser with a power of

100 W. The specimen of the IRG29 glass 3705 collapsed after 600 s of testing, threshold power amounted up to 24 W/cm². The specimen of the glass 3700 withstood radiation equal to 30 W/cm² for 300 s (Table 17). In Table 21 are the results of studies of samples of the glass IRG34 weighing 5 to 15 kg, taken from industrial meltings. Samples of melt 1156 were obtained in consequence of simultaneous synthesis in three founding vessels with 5 kg of glass in each of them at separate preparation of charge out of raw materials of different quality. Sample 3565 was performed in violation of the technological regime. As is seen from the table, at practically equal transmission in the absorption band ($\tau_{\lambda 12.8}$ = 12 to 30%) threshold power of the majority of samples decreases as the number of insertions increases. The sizes and nature of these defects in the samples of different foundings are also different. It's typical for many glasses to have small translucent insertions (10 to 30 µm), wrongfully shaped, probably representing microcrystals of oxides. In some samples, for instance, 4911, large (up to 150 µm), dark insertions stemming from incomplete melting are observed. At continuous laser radiation attack on the glasses of complex compositions the collapse in them, as in the glasses of binary systems, is of a thermal nature: the irradiated part of the sample warms up and the temperature gradient appears between the central and peripheral parts. At attainment of the threshold power it causes the disintegration followed by chaotic cracking. One can see that the disintegration threshold mainly depends on the number of insertions and absorbing radiations with which the heating up of glass melt begins. In the samples containing more than 100 insertions in 10 to 15 min after the testing is started, the temperature rises up to 70°C and the collapse threshold is 25 to 30 W/cm². At the testing of glasses with a small number of insertions the samples heated up to 40°C in 10 min at the power of 40 W/cm². In these conditions the samples don't collapse but the distortion of size and shape of the bunch, i.e., defocusing, is observed; and only after 16 min of testing the temperature of the sample had risen up to 70°C and the cracking began.

The we also came to the conclusion that the stones of incomplete melting absorb radiation slightly. For the most part cracking occurs on them and hence $K_{10.6}$ and radial strength almost don't change (sample 4911). On the contrary, the insertions of oxide origin mainly absorb radiations; that's why the sample 3565, in which plenty of small translucent insertions could be seen, collapsed at power of 13 W/cm², almost immediately after it had been placed in the path of a laser beam. Thus, the right choice of the composition and the accurately performed optimal technological process allowed us to synthesize in working conditions the glasses with radial strength in continuous regime 40 W/cm² which collapsed after 10 to 15 min of radiation attack.

The investigation of radial strength of selenium and binary glasses, established that the character of disintegration in impulse regime changes not only with the change of composition, but also as the radiation power increases. Microscopic investigation of the spots on collapsed surfaces of the samples of glasses IRG29 and IRG34 demonstrated that at power close to the threshold, the burn spot represents the accumulation of craters in the form of open bubbles. With increasing radiation power the separated craters unite and one crater with rough edges is formed. Cracks and punctures observed in binary glasses don't appear at the collapse of the complex glasses.

This type of destruction, apparently, stems from the substance evaporation under the influence of the intense flux of rays. The absence of chips is probably connected with the more homogeneous three-dimensional network of these glasses. The volume scanning under the microscope has shown that volumetric disintegration isn't observed not only at the damage of the inlet surface, but also at simultaneous damage of both inlet and outlet surfaces.

The measurements of the surface radial strength of the glasses IRG29 and IRG34 were carried out in two installations: in GOI — fixing the disintegration threshold — visually, and in FIAN — with the help of an oscillograph — by transparency loss (Table 22). It should be noted that at the attack of radiation of prethreshold power neither spark appearance nor loss of transparency is observed and there are no traces of destruction on the surface. The appearance of plasma flare occurs when threshold power is achieved and can be fixed both visually and with the help of an oscillograph. However, as ChG are opaque in the visible spectrum range, only the surface strength can be defined by the "spark method". This method, which fixes the loss of transparency, allowed us to evaluate broadly the volume strength of glasses at radiation focusing by an NaCl lens inside the sample 45 mm thick. An average among the values of threshold power in different points of volume is equal: for the glass IRG29, about 190MW/cm^2; for the glass IRG34, 240MW/cm^2.

TABLE 22
Comparative Values I_{thr}, Obtained in SOI and PhIAS

Glass type	Melting no.	Threshold power of radiation	
		SOI	PhIAS
IRG29	3314	30 ± 6	32 ± 4
	3499	43 ± 11	48 ± 7
IRG34	3411	57 ± 11	58 ± 7
	3^a(lab)	51 ± 9	50 ± 8

As the samples were studied under the microscope in the path of the beam, deformation zones were revealed — approximately round-shaped and more transparent than the major glass melt. We can suppose that threshold power is appropriate to the energy of bonds' breakage and hence momentary glass fusion occurs in the focus point. As the heat conductivity of ChG is low and the radiation force is of short duration, the melting lies within a microregion just in the radiation area. Macrocollapse similar to that in the crystals KRS doesn't take place here.

For both glasses the influence of accuracy of the surface treatment was investigated and for IRG34, the effect of "laser purification" (Tables 23 and 24). One can see from the table that the destruction threshold increases both at improvement of the surface treatment quality and at "laser purification". Also investigated was the resistance of the glass IRG34 to pulsed radiation 15 μm. The breakdown threshold depends on the treatment to a less extent and is equal, on the average, to 2MW/cm^2.

TABLE 23
The Influence of the Finishing Accuracy on I_{thr} of the Glass IRG29

Surface finishing	$I_{thr}(MW/cm^2)$
On velvet, with water suspension Cr_2O_3, NGP	23 ± 4
On tar, with water suspension Cr_2O_3 + diamond paste, NGP	27 ± 4
The same, DGP, removal 15 min	30 ± 6
The same, DGP, removal 30 min	60 ± 8

TABLE 24
The Influence of Surface Finishing and "Laser Cleaning" on I_{thr} of the Glass IRG34

I_{thr} (MW/cm²)			
NGP	DGP	Laser cleaning I_{thr} (MW/cm²)	I_{dec}
43 ± 11	57 ± 9	$30 \to 33 \to 36 \to 40 \to 64 \to 70$	70

In consequence of the performed investigation it was established that radial strength of ChG at the attack of CO_2-laser radiations in continuous regime at the diameter of a bunch 1 to 2 mm is equal to 0.6 to 0.8KW/cm², which allows one to use them in lasers of low power, mostly when large-size articles (up to 370 mm) are needed.

Threshold power of the surface destruction of the glass IRG34 for lasers pulsers of 0.1 μs is equal to 50 to 60MW/cm², which is commensurate with the radial strength of the majority of optical crystals.

As glass is mechanically more rigid and much less toxic than KRS-5, it is produced in considerably greater amounts than ZnSe, optical ceramics, and GaAs, and is also much cheaper than the majority of crystalline materials, it is advisable to use it as passive elements in devices with impulse lasers.

3.7 CHALCOGENIDE GLASSES FOR FIBER OPTICS[28]

In recent years numerous attempts have been made to apply ChG as fiber optical elements for the devices working in the third atmospheric window, because there are no other commercial materials satisfying this need. Waveguides may be made of ChG by both "fillet-pipe" and spinnereting methods. This glasses exhibit good transparency and low own-optical losses. According to ratings, carried out by Dianov and others, of extrapolation of electron and phonon absorption into the high transparency region — taking into account the losses from Rayleigh scattering — minimum of own losses in ChG is observed in the region 4 to 6 mm and is equal to

0.1 to 0.01 dB/km.[184] The losses in real fibers are, however, determined from the presence of extrinsic absorption and also of imperfections of various kinds; according to the data by Bagrov and others their amount in glasses based on As_2Se_3 is equal to 1 dB/m.[185] The causes of extrinsic absorption bands and their position in the glass spectra are analyzed in detail in Chapter 2. It was estimated that theoretical losses of 10 dB/km may be achieved in fibers of glasses if the content of oxygen, hydrogen, and carbon admixtures not greater than 10^{-6} to 10^{-7} mass%. However, the technical level of commercial production of extra pure substances, and also control over the content of admixtures can hardly make it difficult to achieve such little losses.

The latest works are devoted mainly to establishing the composition of light admixtures and to quantitative determination of their content in initial materials and glasses, and also to the influence of glass purity on optical losses.[184–188]

We may connect to this series of works the review by Devyatikh and Churbanov,[183] dedicated to defining trace contaminants in chalcogenes, and also the review articles by a large group of authors, including the above-mentioned ones, where the highly pure ChG for fiber optics are described.[150] This group of authors managed to obtain fibers based on As_2Se_3 in fluoroplastic cladding with the losses less than 1 dB/m in the wavelength interval 3.8 to 4.5 μm.

Katsuyama and others studied optical transparency of fibers made of the Ge–Se glass.[189] It was demonstrated that introducing antimony into the glass $Ge_{20}Se_{80}$ reduced the losses at the wavelength 10.6 μm from 8 to 3 dB/m. On the basis of the glass As–Ge–Se–Te, worked on by Savage,[146] Parant and others obtained fibers with 5 to 15 dB/m losses at $\lambda = 10.6$ μm by the method "fillet-pipe" in argon atmosphere having elaborated the original way of fillet preparation.[190] All of these results were obtained on glasses synthesized in laboratory conditions.

Hilton offers the commercial glass pair for fibers' drawing. The glasses prepared by the method, elaborated by the author, have the following absorption coefficients: AMTIR-1 (composition As–Ge–Se) — 7×10^{-3} cm^{-1} ($\lambda = 7$ to 9 μm) and 1×10^{-2} cm^{-1} ($\lambda = 10.6$ μm); AMTIR-2 (composition As_2Se_3) — 1.4×10^{-3} cm^{-1} ($\lambda = 7$ to 9 μm) and 3.5×10^{-3} cm^{-1} ($\lambda = 10.6$ μm). Using AMTIR-1 as a coating and AMTIR-2 as a core, the waveguides were prepared with the losses from 0.6 dB/m ($\lambda = 7$ to 9 μm) to 1.5 dB/m ($\lambda = 10.6$ μm), which is best for commercial glasses known from the literature in the 1980s.[181]

We set ourselves the task of creating the glass couple "core-cladding" with the absorption coefficient of the core glass (light-guiding vein) equal to 1 to 3×10^{-3} cm^{-1} at $\lambda = 10.6$ μm, suitable for use in fiber manufacturing. To create glasses for lasers we generally have to select the compositions with the highest thermal characteristics, but it is not necessary in fiber optics, because, as a rule, they don't experience much optical energy. That's why the elements of the IV group of the periodic system may not be introduced here and the glass compositions get considerably complicated.

Rather high requirements are imposed, however, upon couples of the glasses for fiber.

1. To provide high aperture of waveguides the refractive index of the core must be higher than that of the cladding. The aperture number determined by the formula $A = \sqrt{n_c^2 - n_{cl}^2}$ must achieve its maximum.

2. To provide fiber strength the CTE of the core must be equal to that of the cladding or slightly higher.

3. To provide joint simultaneous drawing of the glass couple the viscosity of the core glass must be slightly higher than that of the cladding glass in the interval of temperatures of fiber manufacturing.

4. At fiber production diffusion processes mustn't occur between the glasses of the core and the cladding, leading to the change of composition in the contact layer, because the energy losses may take place therewith due to disorder of the absolute inside reflection.

5. As a part of energy is transmitted through the cladding, it is desirable to reduce the losses of the core and the cladding to have equal transparency in the working range of wavelengths. As mentioned above, the absorption coefficient of both glasses, especially that of the core, should be as low as possible.

Selenium glasses have the best transparency in the long-wave spectrum range, but their refractive index can't be lower than 2.5. Hence, to hold the transparency of the couple and, at the same time, to provide high aperture, one must increase the refractive index of the core glass to its maximum. In this purpose, we replaced a part of selenium with tellurium, based on the system As–Se. The glasses in the As–Se–Te system were studied quite thoroughly and the least crystallizing composition there were obtained along the cut AsSe–AsTe.

Further investigations have demonstrated that for manufacturing the most overlooked is the glass standardized as AST-1. This glass was investigated to study the dependence of optical losses on the quality of initial materials. It is known that in commercially provided reagents, even chemically pure without metal and nonmetal admixtures, there are inevitably present SeO_2, As_2O_3, and TeO_2; hence, all these elements appear to be sources of oxygen. It is quite easy to rid selenium of oxide through its fusion in dynamic evacuation regime at a temperature of about 350°C. The time of warming up is from one to several hours, depending on the quantity of selenium. At the same time it gets rid of moisture by the reaction

$$3Se + 2H_2O \rightarrow 2H_2Se + SeO_2$$

Commercial arsenic contains oxides As_2O_3 and As_2O_5. Upon heating under the same conditions pentoxide disproportionates on As_2O_3 and O_2 and trioxide, as well as oxygen, is easily sublimated and removed by vapor evacuation.

Technical-quality tellurium isn't that easy to purify. Dioxide presence in it manifests itself upon warming the metal in the evacuated quartz ampul at T = 400°C for 3 to 4 h. TeO_2 appears on the surface of metal and the walls of the ampul as a dirty white scale. Its particles are very firmly attached to the surface and are hard to remove mechanically. The stability of TeO_2 and the high temperature of phase transitions don't enable one to remove it from metal by heating. Distillation of tellurium is also not successful due to comparable pressure values of metal and its oxide. Besides, tellurium condensates after distillation in the form of spiny crystals with a very mature surface, which even at short-period contacts with the atmosphere reacts intensively with both oxygen and moisture. In this case tellurium dioxide is also formed by the reaction $Te + 2H_2O \rightarrow TeO_2 + 2H_2$. We have advanced a method of tellurium purification of TeO_2 based on the reducing tendency of oxide. Thus, it is known that TeO_2 is easily reducible to free metal, for example, by hydrogen:

$$2H_2 + TeO_2 \rightarrow Te + 2H_2O$$

Thermodynamic calculations have demonstrated that the reducibility reaction at TeO_2 interaction with arsenic must also go: $4As + 3TeO_2 \rightarrow 2As_2O_3 + 3Te \quad Q = -337.6$ kJ/mol.

The kinetic barrier of the reaction lies lower than 250°C. Hence, the warming up of tellurium contaminated with TeO_2, with arsenic at T = 300 to 400°C, must lead to formation of As_2O_3 vapors as the reaction goes in the gas phase, and easily volatilized arsenic oxide is removed during the carrying out of the reaction in the dynamic evacuation regime.

The surface of tellurium purified in this way is not as active and reactive as at distillation of the metal, which enables one to work with it in a normal atmosphere.

Transmission spectra of the glasses of the system As–Se–Te melted of tellurium, purified by this method (curve 3), and technical-grade tellurium of different contamination degrees (curves 1 and 2) are shown in Figure 49. The absorption coefficient of pure glasses for $\lambda = 10.6$ μm steadily lies within the limits 7 to 8×10^{-3} cm^{-1}, whereas the glasses melted with unpurified tellurium had $K_{10.6} = 3$ to 7×10^{-2} cm^{-1}. We should note that this method of tellurium purification enables us to use cheap raw materials. However, this level of absorption, sufficient for normal optics, is still too high for fiber products.

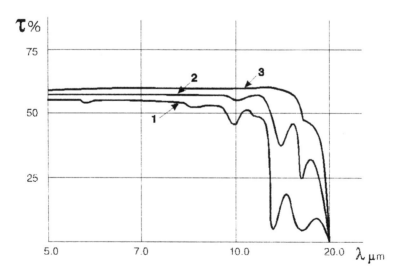

FIGURE 49. Transmission spectra of the glass As–Se–Te with tellurium of different purity.

Tellurium of the necessary quality is obtained by the method of directed crystallization described by Chulzhanov.[191] The complex tellurium purification offered by the author includes distillation, directed crystallization, reducing fusion, and other operations, providing obtaining the product with total content of admixtures of 21 metals $<1 \times 10^{-5}$ mass%, gaseous admixtures CO_2, H_2O, and H_2 $<5 \times 10^{-5}$ mass%. Arsenic for the extra pure glasses must also be purified through directed crystallization and evacuated distillation. The allowable content of admixtures is 10^{-5} to 10^{-7} mass%, namely, of oxygen — 6×10^{-6} mass%; carbon — 2×10^{-6} mass%; of gaseous CO_2, H_2O, and H_2 — $<5 \times 10^{-5}$ mass%. Selenium is purified by the method proposed by Pazukhin and Fisher through thermal treatment at dynamic evacuation. The temperature in the precipitation zone influences significantly the quality of purification. The most effective evaporation temperature, according to the data by the author of the method, is equal to 400°C, and condensation temperature to 190°C.[192] The application of raw materials of such purity primarily aims at the performing of synthesis in the conditions allowing it to hold low level of admixtures. The following is necessary:

- To make founding vessels of high-quality melted quartz; to put them through thorough chemical purification before synthesis; to wash them out by distilled water with less than 0.1 mg/l oxidizability and the contents of heavy metals $<1 \times 10^{-6}$ mass%
- To perform the calculation of vessels at ~1000°C in the regime of dynamic evacuation
- To prepare initial substances and to compose the charge in bays in the atmosphere of dry argon; argon content 95 vol%, oxygen admixtures not more than 1 ppm (To define the optimal interval, necessary to prepare the bay to work, the composition of the atmosphere of the bay should be controlled.)

At blast cleaning of the bay simultaneous extruding and dilution of the air by argon take place. The volume part of argon in the bay (C) is calculated by the equation:

$$C = 1 - e^{-M}$$

where $M = RT/V$ is shortness of the blast cleaning, i.e., the ratio between the total volume of argon delivered into the bay and the volume of the bay; R is consumption of argon; T is time; V is volume of the bay.

The composition of the atmosphere of the bays is defined by the diffusion chromatographic method. It was established that the content of oxygen equal to 1 vol% in the bay is achieved at $M = 3$, i.e., at three-stage blast cleaning.

- To evacuate the charge until discharging 10^{-5} to 10^{-6} mm mercury
- To elaborate time-temperature synthesis regime for each glass type, taking into account separately the granulometric charge composition and the volume of the product being melted

With these observed recommendations the glass ASTl was melted in industrial conditions. The absorption coefficient, measured by the laser calorimetric method, was equal to 1×10^{-3} cm^{-1} ($\lambda = 10.6$ μm) and 2×10^{-3} ($\lambda = 5$ to 6 μm). It should be noted that if the absorption coefficient of pure selenium glasses at $\lambda = 5$ to 6 μm is lower than at $\lambda = 10.6$ μm, in tellurium glasses the proportion is inverse. It can be explained apparently by the presence of an asymmetrical absorption band with the center 5 μm due to H–Te bonds in the zone of CO_2-laser radiation. In this way, the glass ASTl can be used as the core of the fiber-optical pair. The IRG29 glass can serve as a cover. Data of these glasses' properties is given in Table 25.

It's of interest that in the 1990s "Amorphous Materials Inc." company advertised the glass AsSeTe as the material for optical fiber. The properties and quality of the glass were not mentioned. However, if to take into account that in the system As–Se–Te there are compositions of noncrystallizing glasses adaptable for purification by distillation, we can consider the glass of the "AMT" company optically pure and applicable as a light-guide.

To work in a shorter wavelength region (to 10 μm), trisulfur arsenic may be used as the core — it's the only noncrystallizing chalcogenide. However, to obtain glass with low losses its supplementary purification after synthesis is necessary.

Without analyzing the known methods for As_2S_3 purification,[84,193,194] we'll describe the technical process elaborated by us to obtain extremely pure vitreous sulfide, which includes the synthesis of glass, supplementary purification, and the further homogenizing of the product. This process is based on the most effective method of supplementary glass cleaning by vacuum distillation. To synthesize the glass, sulfur and arsenic, deeply cleansed of light admixtures, are used. In this case vacuum distillation allows one to additionally purify the produced glass of infusible substances which form heterophase insertions (for example, SiO_2).

The purification of As_2S_3 is obtained if arsenic is used with the content of admixtures on the level 10^{-5} to 10^{-7} mass%, including oxygen on the level 6×10^{-6} mass%, carbon 2×10^{-6} mass%, and sulfur with the content of bitumen on the level (5 to 10) 10^{-5} mass%. Preparation of charge is carried out in bays in argon atmosphere. After washing the founding vessel is calcinated at 1000°C and evacuation is carried out at the pressure 10^{-5} to 10^{-6} mm mercury. The synthesized glass is cleaned by the method of vacuum distillation and after that the condensate is homogenized. The absorption coefficient of the glass, obtained by this method in the spectrum range 5 to 6 μm, appeared to be 9×10^{-4} cm^{-1}

3.8 INFRARED OPTICAL GLUES AND IMMERSION MEDIUMS FROM CHALCOGENIDE GLASSES[29]

Chalcogenide halogen-containing glasses, due to their low melting point, enable one to develop compositions, adaptable for agglutination of different IR materials and also as immersion mediums. Cases are known when ChG were used for these purposes, however, the melting temperatures of the compositions are rather high (250 to 300°C) and it complicates the technology of their application in components.

TABLE 25
The Principal Properties of the Glasses AST1 and IRG29

Optical properties

Glass type	Transparency region ($\lambda\mu m$)	Maximum transparency (%)	Refractive index ($n_\lambda/\lambda\mu_m$)							Dispersion ($n_{\lambda 1} - n_{\lambda 2}$)		Dispersion coefficient (ν)	
			5	6	7	8	9	10	12	n_5-n_7	$n_{10}-n_{12}$	ν_6	ν_{10}
AST 1	1.5–17	62	3.02	3.018	3.015	3.014	3.012	3.01	3.007	0.005	0.007	404	287
IRG29	1–15.5	65	2.614	2.611	2.609	2.606	2.604	2.601	2.593	0.005	0.013	322	122

Thermal Properties

Glass type	$\alpha_{30-120} \times 10^6$ (°C^{-1})	T_{ds} (°C)	Temperature course of viscosity ($\lg\eta/T$)									
			13	12	11	10	9	8	7	6	5	4
AST 1	20.3	182	140	156	170	186	208	233	261	294	330	376
IRG29	22	210	173	190	207	223	230	257	276	302	332	368

Crystallizability and Density

Glass type	Crystallizability after 6 h of hold time at 150–550°C	Density (g/cm³)
AST 1	0, doesn't crystallize	4.925
IRG29	0, doesn't crystallize	4.74

Thus, Kodak company used fusion in the 1960s, consisting of sulfur, selenium, and arsenic trisulfide as the optical glue, transparent to 13 μm.[195] The well-known U.S. patent exists in which the glass consisting of selenium, sulfur, tellurium, and arsenic is put forward to create an immersion layer on the surface of a germanium bolometer.[196] The layers of this glass are applied by the method of vacuum evaporating with the subsequent agglomeration. Thermoplastic glues and immersion mediums have certain advantages: they can connect large-size details of various configurations, therewith the technical process of agglutination is simplified. The most difficult is the creation of glass compositions with low melting temperature, and, at the same time, with little reactivity of the fusion para to atmosphere oxygen to hold transparency at agglutination. The gluing layer must also withstand the temperature gradient ±60°C and provide agglutination temperature not over 170°C. Therewith it must keep vitreous with no signs of crystallization for at least 6 h, i.e., during agglutination and the subsequent annealing of components.

In the systems where light-diode and radiation detector are made of highly refracting materials (n ~ 3.0 to 3.2) the glasses must have extremely high refractive index to be used as immersion medium (vehicle). Therewith the refraction losses on the border of two mediums decrease. Besides, required is their satisfactory adhesion to silicon and gallium arsenide, chemical stability, inactivity in the point of contact, and great specific resistance (not less than 10^9 Ω/cm). The instruments made of these materials must retain efficiency for many years.

As far as is known, Fischer was the first, in 1969, to use ChG with halogens as immersion medium.[197] He used the glass As–S–Br to enhance efficiency of light diodes on gallium arsenide. The research of Ukranian scientists Turyanitra, Khiminetz, Dovgashey, and others demonstrated the possibility of applying the technique of halogen-containing ChG in semiconductors, optic electronics for integral schemes, information recording, etc.[198,199]

Based on the literature data and our investigations in an effort to create commercial glasses, systems containing arsenic, sulfur, and selenium with the additions of thallium, antimony, and iodine have been studied. One of the most important criteria which determines the possibility of practical application is, as in all other cases, crystallizability.

The glasses in the system As–Se–Tl–I and also the glasses with tellurium have rather high crystallizing tendencies. All attempts to reduce it by adding heavy elements of the IV group ended in enhancement of infusibility and decrease of transparency in the near part of the spectrum, which is also undesirable. The glasses, not crystallizing in a wide temperature interval at 6 h of endurance, were obtained in the systems As–Sb–S–I and As–Sb–Ge–Se–I. One of them, the commercially manufactured one, is marked as TKS-1. The transparency of TKS-1 in the depth of a gluing layer 22 μm, and also of the samples of germanium and the glass IRG30 agglutinated with its help, is shown in Figure 50. One can see that the transmission of the agglutinate samples at equality of the refractive indices of glue and sublayers changes little (curves 4 and 5). Similar results were obtained at agglutination of the samples of cesium iodide, antimonous indium, and IRG25. TKS-1 application is possible at temperatures ±60°C. At the same time in some cases it is required that the working temperature interval is –60 to +120°C. For these purposes two glasses were worked out: BV_1 and BV_2. In Table 26 properties of these glasses are given. It can be seen from the table that at the equality of thermal characteristics the glasses BV_1 and BV_2 have very different refractive indices ($n_{2.2}$ = 2.306 and 2.603, respectively). Hence, although the glass BV_2 is a bit less technological, their application is preferable when it's necessary to connect high-refracting media. The agglutination temperature of the glass BV_1 = 205 to 215°C; BV_2 = 260 to 280°C. Dielectric properties of the glasses, measured on the wavelength 3 cm, are given in Table 26.

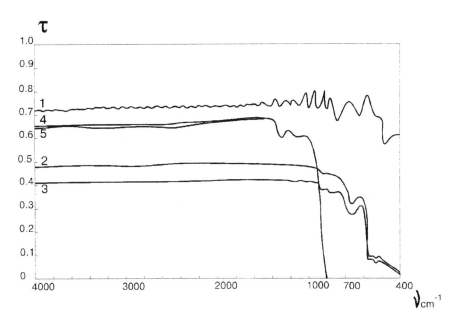

FIGURE 50. Transmission of the samples, glued by TKS-1: (1) Film TKS-1, 22 μm thick; (2) germanium 6 mm thick; (3) glued germanium 3 + 3 mm thick; (4) IRG30 10 mm thick; (5) glued IRG30 5 + 5 mm thick.

3.9 THERMOPLASTIC GLASSES FOR OPTICAL ELEMENTS OF IR SPECTROSCOPY OF BROKEN ABSOLUTE INTERNAL REFLECTION[30]

Spectroscopy of Broken Absolute Internal Reflection (BAIR) is used in cases when we can't apply the methods of transmission spectroscopy, i.e., in the absorption range of the objects under study and also at analytical measurements.

Optical element BAIR is formed out of the material with the refractive index n_1, which is higher than the refractive index of the sample n_2. The internal reflection occurs on the interface of two mediums where $n_1 > n_2$ at angles of incidence over the critical one. Therewith the shaft penetrates into the less optically dense medium on the depth in the region of wavelength.

TABLE 26
The Properties of IR Optical Glues

Glass type	Transparency limits in the thickness of the gluing layer (~20 μm)λ_{thr}	Agglutination temperature (°C)	Working temperature range	T_{ds} (°C)	$\alpha_{30-100} \times 10^6$ (°C^{-1})	ρ (g/cm³)	Refractive index/$\lambda_{\mu m}$				
							0.65	1.8	2.2	2.6	
TKS-1	0.6–25 μm	160–170	−60 + 60	62	42	—	2.348	2.221	2.214	2.208	
BV-1	0.6–25 μm	205–215	−60 + 120	126	38	3.34	2.45	2.309	2.306	2.299	
BV-2	0.75–25 μm	260–280	−60 + 120	125	34	4.66	—	2.615	2.603	2.599	

If an optical element is made of hard material, the necessary contact with a sample can be achieved after thorough polishing of an object and an element in the appropriate color with subsequent mechanical pressing down.

In the classical technique for the elements BAIR, quartz, silver chloride, KRS-5, silicon, and germanium are used the most. Optical treatment of these elements is laborious, sometimes toxic (KRS-5); at the same time their life is short: they can withstand only several measurements. While lying on optical contact it's difficult to achieve full contiguity with a sample and one has to use a press. The best contact of an element with an object is achieved when fluid elements are applied. Unfortunately, there aren't many refracting IR-transparent liquids. Carbon disulfide ($n_d = 1.62$) and methylene iodide ($n_d = 1.74$) are most commonly used for these purposes. One example of successful application of high-refracting liquid (n = 2.0) is the fluid glass from the system As–S–Br, described by Saidov.[200] However, for any liquid cuvettes are needed and their walls must be also well transparent in the IR region; fluid spectroscopy doesn't enable one to investigate large-size objects.

The application of thermoplastic glasses for the preparation of optical elements BAIR is of considerable promise, i.e., therewith the drawbacks typical of hard and fluid mediums can be avoided.

Neither an optical transparent cuvette nor mechanical treatment of an element and an object is needed, which significantly widens the possibilities of the technique. The special requirements are, however, imposed on the elements of BAIR. Along with the high refractive index at the maximum wide transparency range, they must possess good adhesion to objects and be chemically stable. Their thermal properties and crystallizability must allow molding at low temperatures with preservation of optical homogeneity. Low-crystallizing glasses in the system As–Sb–S–I have the molding temperature, i.e., they transfer into thermoplastic state from fusion at approximately 110 to 120°C and provide optical contact without separation with the sample at room temperature. The transparency region of such glasses lies in the limits 0.6 to 10 μm at a thickness of 2 to 5 mm, but in working depths (10 to 20 mm) they have significant absorption bands. Their refractive index is about 2.2.

The glasses in the system As–Se–I are more transparent in the long-wavelength part of the spectrum and have n about 2.4. The transparency region of such glasses is equal to 1 to 18 μm in depth up to 10 mm. There are compositions among them with the molding temperature 50 to 100°C. Hence the system As–Se–I became the basis for elaboration of commercial glasses. Complication of the composition to reduce the crystallizing tendency went along the line of partial replacement of arsenic with antimony and of selenium with tellurium. As a result the glasses were obtained which don't crystallize during heat treatment and have the refractive index close to KRS-5. The molding temperature for different compositions changes from 30 to 40°C to 80 to 100°C. One such glasses denoted as IRG35 is produced in 5 kg quantities in one melting.

Values of the refractive index are given in Table 33 in Chapter 4. In Figure 51 (curves 4 and 5) BAIR spectra of fluoroplastic and quartz are shown, obtained with the help of BAIR attachment, elaborated by Zolotarev, with optical elements made of the glass IRG35.[201] Apart from the spectroscopy BAIR IRG35 can be applied as a gluing layer for the connection of IR optical details. As an example we refer to the transmission of two plates KRS-5 connected through the layer of the glass IRG35 (Figure 51, curves 2 and 3). IRG35 is convenient to use as a temporary optical contact at the measurement of physical characteristics of the articles not put through the optical treatment — for instance, of the surfaces of chips or of fibers' ends. Another application of IRG35 is possible in measurements of IR spectra of unrefined crystals, powders or of substances with the refractive index close to that of IRG35, sealed inside the glass. In Figure 52 examples of IRG35 application for the spectroscopy BAIR are demonstrated.

3.10 INFRARED FILTERS BASED ON CHALCOGENIDE GLASSES[31]

The function of light filters in various optical instruments is a matter of common knowledge, and modern IR techniques require filters which can cut off visible and nearest-IR spectral ranges to abolish interference caused by the sun's radiation. In the devices for thermal vision and photographing through the atmosphere thickness, namely, in space instruments, band filters are required eliminating the second or the third atmospheric windows. Therewith transparency in the long-wavelength region is also undesirable because of atmospheric interference.

Crystalline film and interference light filters, used for these purposes, have drawbacks which limit their application. Glass light filters are of great interest, for they display certain advantages. Among them are, first and foremost, the following:

1. Greater simplicity of manufacturing. Thus, manufacturing of interference light filters for IR-spectrum range requires the applying of several dozens of layers on the substrate.
2. Greater safety and longevity in comparison with film filters, namely:

   a. Higher heat resistance due to the exclusion of films' expoliation after thermal shock
   b. Higher chemical stability, for the films contain chemically unstable elements, for example, tellurium
   c. Independence on the surface state — scratches influence spectral characteristics of the films
   d. Immutability of the characteristics in the course of time, as deterioration of the filters, owing to the films' recrystallization, is excluded

3. Selection of advantages of the spectral characteristics

   a. Completeness of the cutoff of a short-wave region
   b. Spectral characteristics' independence on the angle of light incidence

For light filters with selective absorption of ultraviolet, visible, and short-range IR radiations colored optical glasses are used. Light filters with selective absorption of radiations of middle- and far-IR band may be called colored IR glasses. The

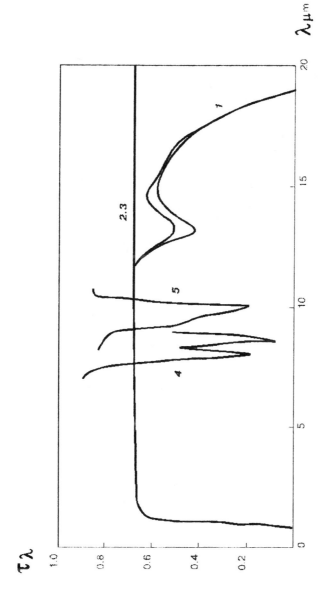

FIGURE 51. Transmission of the glass IRG35: (1) samples 10 and 20 mm thick; (2) film 20 to 30 μm thick; (3) glued crystal KRS-5; (4) spectrum of the broken total internal reflection in fluoroplastic; (5) spectrum of the broken total internal reflection in quartz.

FIGURE 52. Examples of IRG35 application as the element of BAIR spectroscopy.

absorption bands' shift is obtained in two ways: either by composition change through the introduction of the components, which form color centers caused by electron transitions of various types, or by formation of colored centers at repeated heat-treatment in the form of colloid particles of various sizes in the same glass.

When creating IR light filters based on ChG, we used both these possibilities. Recall that glasses of the majority of compositions in the systems As–Ge–Pb–Se and As–Ge–Pb–Sn–Se contain additional lead selenide which cause scattering and absorption of radiation, which can't be revealed under the microscope due to the high dispersion degree and comparatively little amount. The distinctive property of such glasses is in quite strong dependence of the position of short-wavelength transmission edge (λ_e) on the heat-treatment regime. The longer the glass is kept in the temperature range excluding T_{ds}, by 50 to 150°C, the further λ_e shifts to the IR region. To find out the dependence of the position of transmission edge on the thermal past of the glass, the initially noncrystallizing compositions were chosen and the synthesis of glasses was carried out in spectral conditions. The charge was put into the cylindric quartz ampul Ø20 to 25 mm, about 250 mm long. After melting the ampul was taken out of the furnace, cooled in the air in horizontal position, then held for 6 h in the specially graduated gradient furnace and cooled again to room temperature. After that the ampul was unsealed and the band of glass was taken out of it and cut into samples about 10 mm long. Both sides were polished, after which their spectral characteristics were taken.

As the interval of heat changed from sample to sample, the edge wavelength-of-temperature chart could be plotted on the basis of the received data (Figure 53). As it follows from the picture, the edge wavelength up to T = 320°C keeps practically

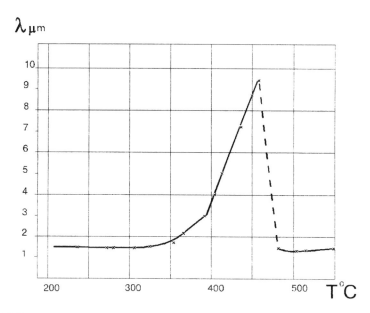

FIGURE 53. The edge waveline dependence on the temperature of IRG27 at 6 h holding time. The samples are 2 mm thick.

constant, then it begins to shift slowly to the long-wavelength region; therewith its steepness as well as the maximum transmission don't change. In the temperature range 390 to 460°C the abrupt shift of the transmission edge to the long-wavelength part of the spectrum is observed, followed by the fall of the threshold steepness and the transmission maximum. At T = 460 to 470°C the curve experiences the rupture, the value λ_e returns to the initial one, and from here on it is kept constant. It should be considered that the shift of transparency edge is caused by the growth of the sizes of PbSe particles and hence this process is identical to that of colored glasses produced by colloid coloring. The time interval, necessary for the creation of the filter with the prescribed position of the transparency edge, was established after thorough examination of the process of PbSe particle growth held at a certain temperature. This process can be likened to the heterogeneous reaction in the glass melt. Then, according to the Nernst theory of heterogeneous processes, the growth rate of PbSe nuclei is determined by the diffusion of lead-containing structural units through the transmission layer, in which their concentration changes linearly from C_0 in the glass volume to zero near the surface of a nucleus. In terms of fixed diffusion this rate will be equal to

$$R = (D/X) \times C_0$$

where D is the diffusity coefficient and X is the thickness of the transition layer. The amount of lead selenide, liberated at the nucleus in a definite time dt, is calculated from the formula:

$$dm = (D/X)C_0 \times S \times dt$$

where S is size of the surface through which the diffusion goes.

Considering this surface to represent the sphere with a radius equal to the thickness of a transition layer X (one can disregard its change in the process of a nucleus growth), we rewrite the formula (13) as follows:

$$4\pi r^2 dr \times \rho = (D/X)C_0\, 4\pi X^2 dt$$

where ρ is substance density of a particle (PbSe) and r is radius of a particle. Hence

$$\rho r^2 dr = D \times C_0 \times X dt$$

Let us integrate the term (15) from the starting point of the heat treatment t_0 to the moment of time t, indicating by r_0 the radius of a particle before the heat treatment begins. We obtain

$$\rho \int_{r_0}^{r} r^2 dr = DC_0 X \int_{t_0}^{t} dt$$

$$\frac{1}{3}\rho(r^3 - r_0^3) = DC_0 X(t - t_0)$$

and hence

$$r^3 = \frac{3DC_0 X}{\rho}(t - t_0) + r_0^3$$

In this way, cubic radius of PbSe particles must be in linear dependence on the heat treatment time interval.

It is known that in the case of the presence of the second-phase particles in transparent medium, the maximum value of dispersion coefficient is observed in the quite certain relation between the radii of particles and wavelength, namely,

$$2\pi r/\lambda = 2$$

Then λ_e at constant thickness of samples to a first approximation must be proportional to linear dimensions of particles, hence its cube must depend linearly on the heat treatment time, i.e.,

$$\lambda_e^3 = \lambda_0^3 + A \times t$$

where λ_0^3 is position λ_e for the glass without heat treatment (const.).

Experimental testing showed that when the heat treatment time isn't too long, the linear dependence is really obeyed. This enables one to obtain IR filters with the transparency edge prescribed beforehand. The adaptability of glass for filter making through the setting is apparently determined by two reasons. First, the glass must be easily susceptible to setting, i.e., enough lead-containing structures should be present in it. Second, it must have low tendency to crystallization of the principal vitreous matrix. This exact crystallization, which can start during the setting, explains the abrupt deterioration of transmission in the temperature range 390 to 460°C. The temperature equal to 460 to 470°C is apparently that of liquidus, over which all the crystals transfer into melt, including PbSe. The glass, chosen for commercial management, was denoted as IRG27. On its base filters with any prescribed position of λ_e in wavelength intervals 1.3 to 3.5 μm are manufactured. The steepness of transmission edge is determined by the width of spectrum range where the transmission increase from 0.1 to 0.9 τ_{max} changes from 0.6 to 0.8 μm (for filters with the edge to 1.5 μm) to 1.4 to 1.6 μm (for filters with the edge about 3.5 μm).

The dependence of the position of the short-wavelength transmission edge from the temperature and setting time is presented in Figure 54. The most favorable heat treatment temperature is equal to 350°C.

It is necessary to mention that the constant term λ_0^3 in formula (19) the Figure 54 appropriates to the transmission edge at the laboratory regime of glass-cooling after the setting. After the change of this regime the angular coefficient A in formula (19) keeps unchanged, but the value λ_0^3 is determined by the number and size of particles PbSe, which appear in glass in the process of cooling from the molling temperature to the setting one. In particular, it depends on heat capacity and heat insulation of equipment used at molling and setting in industrial conditions.

Spectral curves of the set of commercially manufactured light-filters are given in Figure 55. Transparency of glass in the long-wavelength spectrum range is limited about 16 μm. Filter transmission may be enhanced up to 90 to 95% through clarification. In light-filter manufacturing, the dependence of the position of short-wavelength transmission edge on the thickness of samples is of great importance. Thus for a 1 mm increase of a sample's thickness, the edge is offset by 0.1 μm. This dependence is used for the final correction of the position of transmission edge. The most favorable filter's thickness is 2 to 3 mm.

The T_{ds} of IRG27 is equal to 270°C; CTE $17.7 \times 10^{-6}/°C^{-1}$; microhardness 190 kg/mm², density 4.89 g/cm³.

Filters for the third atmospheric window must have the transmission edge beyond 6 to 7 μm. In glasses of the IRG27 type it appears to be impossible to shift λ_e further than 3.5 to 3.7 μm without crystallization of matrix neither by composition change nor by setting. Stable-enough glasses with reproducible transmission edge and suitable for setting were obtained in the systems containing As–Ge–Pb–Se, at the replacement of a part of selenium with tellurium.

Crystallizability of glasses in the temperature range of 300 to 550 to 700°C was measured in a gradient furnace at various holding times. Temperature dependence on crystallizability, spectral transmission, and λ_e is given in Figure 56a for three glasses at 2-h exposition. One can see that in all glasses at temperature interval of 280 to 450°C complete crystallization takes place, which is visually estimated

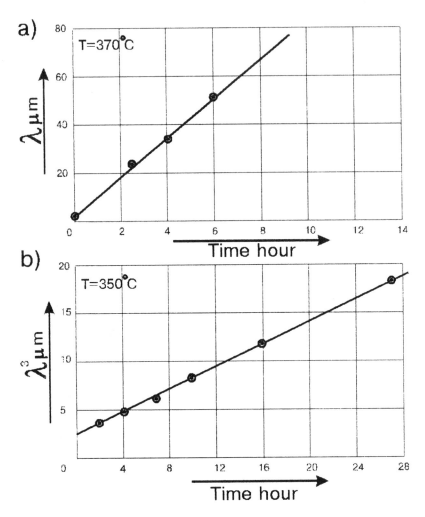

FIGURE 54. The dependence of the edge wavelength cube on the holding time for IRG27: (a) at T = 370°C; (b) at T = 350°C.

according to the type of samples' chip and is proven by the transmission decreasing to zero.

In the range of temperatures adjacent to crystallization edge, shifting of λ_e to longer wavelengths occurs. Glasses 2 and 3 provide shifting up to 4 to 5 μm; glass 1 up to 6 to 7 μm. Based on this experiment we chose glass 1 for further investigation. We tested the influence of heat treatment on λ_e on this glass with the proviso that the maximum transmission value remains unchanged. In Figure 56b one can see that as the holding time in gradient furnace reduces, crystallization region decreases from 250 to 450°C (at 2-h exposition) to 330 to 430°C (at 15-min exposition). Nevertheless, crystallizability remains rather high and at enhancement of the hold time the

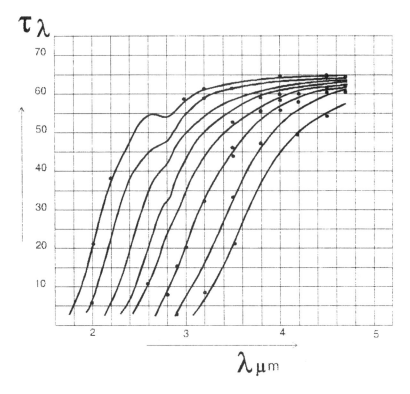

FIGURE 55. Position of the transmission edge in the industrially manufactured set of filters made of glass IRG27. The thickness is 3 mm.

glass becomes opaque. Let us note that in consequence of small sizes of the second-phase particles they remain unnoticed at visual examination.

For the final establishment of the nature of shifting of the short-wavelength transmission edge in ChG, containing heavy elements, electron-microscopic photographs were taken from the heat-treated samples of glass 1. For this, new chips of the glass had been pickled 1 N by NaOH solution for 5 to 20 s; after that they were washed in ethanol and dried out. From the obtained samples replicas were taken, created by simultaneous sputter of coal and platinum which were examined under the electron microscope UEMV-100K at 10,000 to 40,000× magnification.

In all the samples drop-shaped nonuniformities were discovered, which is evident of their liquating origin (Figure 57). As the temperature rises, the dimensions of drops increase pass the maximum, and then decrease. Compositions of phases in the system As–Ge(Sn,Pb)–Se(Te) comply with glass with excess of selenium (matrix) and polycrystals, enriched with lead and tin (drops). Dimensions of liquating drops are appropriate to the wavelength of the radiation, which enables one to use such two-phase materials as optical IR filters.

Melting and manufacturing of glasses containing heavy elements were carried out in the appropriate regimes with quenching from 500 to 600°C to room temperature; annealing of blanks at the rate of 5 to 10°C/h.

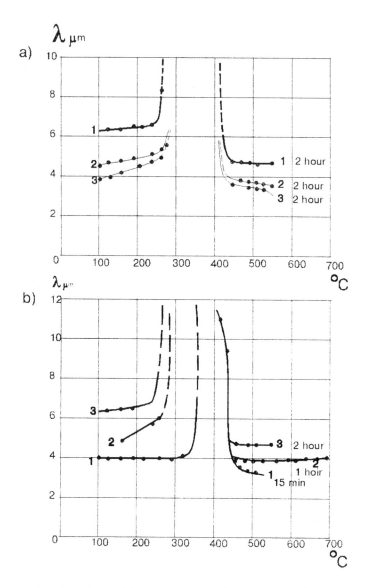

FIGURE 56. The dependency of the edge wavelength in the filter IRG33 on the temperature and time of heat treatment.

Prominence was given to the technique of thermal treatment. In consequence of the detailed examination of the influence of temperature and crystallization time over λ_e position, it was established that even slight deviations of the heat treatment regime and also thermal prehistory of the samples cause considerable oscillations of light transmission and of the position of transmission edge. We managed to stabilize heat treatment regime, having combined it with the molling process. After synthesis and annealing the glass is cut into small plates of certain weight for molling into the disks

$\overline{\mu m}$

FIGURE 57. Electron replica micrograph of IRG33 glass.

of prescribed size, put into molds, and warmed up to 500°C, i.e., over the liquidus temperature, in the special furnace. After a certain hold time, sufficient for the dissolving of all crystallization nuclei formed at synthesis and for making preparations into specified shapes, they are put into the furnace warmed up to 200°C. At cooling, in the temperature interval 350 to 440°C, the isolation of the dispersion phase occurs in a rather short period of time, insufficient for matrix crystallization. Beginning with 200°C, preparations are put through fine annealing. If the equipment for molling is properly assorted, the cooling time is constant for the prescribed temperature range and it hardly depends on its oscillations within the error of the measuring instrument. It provides high reproducibility λ_e at constancy of light-transmission maximum. At the noted heat-treatment regime the samples with extreme thickness of 2 to 5 mm can be used, because with its enhancement the transmission maximum decreases at considerable shifting of λ_e. Easily reproducible results were obtained for the blanks 30 to 50 mm across, 6 mm thick. The commercial glass filter got the index mark IRG33.

The characteristics of the glass IRG33 are as follows:

- Short-wave edge λ_e, μm 7.5 ± 0.5
- Long-way edge λ_e, μm 16.5
- Transmission at a maximum, τ_{max}% 60 to 64
- Coated sample transmission at a maximum, τ_{max}% 90
- T_{ds} 183 ± 2°C
- $\alpha \times 10^{-6}$ °C^{-1} 22.5
- Microhardness, 10^{-7} Pa 147
- Density ρ, g/cm^3 4.968
- Refractive index in the range $\lambda_{7.0}$ to $\lambda_{14.0}$ = 2.682 to 2.658

The glass-filter IRG33 has a lower transmission ratio than interference filters; when it's necessary to increase the ratio, it can be used as a sublayer. The number of interference layers therewith, naturally, considerably decreases.

As mentioned above, in some instruments the transmission of glasses beyond the third atmospheric window is also undesirable. Filters for 8 to 13.5 μm are necessary. Transmission in the near-IR-spectrum range may be removed with the help of an assortment of interference layers, but the far edge can't be formed due to great reflection losses from films. For this purpose special glass is needed; its transmission in the long-wave range must reproduce as completely as possible the edge of the third atmospheric window. We decided to reduce the transparency of one of the already known IRG in a long-wavelength spectrum range. All the attempts to introduce oxygen as oxides into the charge were of no success: the insertion of 0.01% O_2 wasn't sufficient for complete transparency loss and already 0.03% O_2 reduced transmission throughout the spectrum. Then small sulfur addings were inserted as oripigment into the charge by equimolecular replacement of arsenic and selenium. After sulfur insertion into selenium-containing mediums we obtained the series of glasses with various transparency in the 10 to 18-μm wavelength range. The spectra of these glasses in comparison with IRG24 are given in Figure 58. Glass among them was chosen, standardized by the mark IRG28. Its $T_{ds} = 200°C$, CTE = $22 \times 10^{-6}/°C^{-1}$. On the basis of IRG28 band-pass filters for the 8 to 13-μm range are prepared. A certain decrease of glass transparency in this spectrum range, as well as transmission decrease, due to the interference layer are compensated by clarification of ready filters' surfaces. Integral transmission in a working spectrum range is 60%.

The advantage of the use of IRG28 is in the possibility of manufacturing lens of a necessary thickness (hence, of a necessary diameter) which excludes additional filters' necessity.

3.11 CHALCOGENIDE GLASSES FOR
ACOUSTO-OPTICAL DEFLECTORS[32–34]

Two groups of techniques for optical radiation control are known: electro-optical and acoustic-optical. The acoustic-optical modulation methods are based on acoustic-optical effect involving the change of a substance refractive index under the influence of mechanical stresses. There exists the following correlation between the change of the refractive index of medium Δn and the change of pressure ΔP:

$$\Delta n = \frac{\left(n_0^2 - 1\right)\left(n_0^2 + 2\right)}{6n_0 \rho v^2} \times \Delta \rho$$

where ρ is density of material; V is acoustic velocity in a medium; and n_0 is refractive index of a medium.

The principal parameter which is typical of the effectiveness of acoustic-optical medium is the acoustic figure of merit M_2:

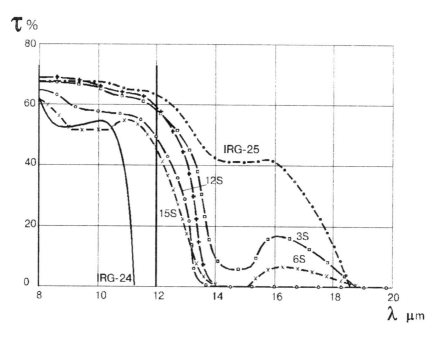

FIGURE 58. Spectra of the glasses IRG25, with additions of sulfur, and IRG24.

$$M_2 = \frac{n^6 p^2}{\rho V^3}$$

where p is photoelastic constant. The usage of materials with the higher M_2 allows one to reduce the driving power necessary for incident radiation deviation.

The limitations imposed on the maximum width of an acoustic bunch give the second M_1 figure of merit criterion:

$$M_1 = \frac{n^7 p^2}{\rho V}$$

The condition of optimal width of light bunch in the interaction region "light-sound" suggests the figure of merit criterion M_3:

$$M_3 = \frac{n^7 p^2}{\rho V^2}$$

In this way the assembly of three parameters — coefficient M_2, ultrasound rate, and ultrasound damping in a material — characterizes in detail a material as an active acoustic-optical medium. All these parameters are a maximum in materials with a high refractive index, high photoelastic constants, low density, and low sound rate.

In the IR range many materials exist with a high acoustic-optical effect, mainly caused by a high refractive index. Among them are the crystals of tellurium, germanium, gallium arsenide, and others. The most effective are the crystals of tellurium ($M_2 \approx 4.400 \times 10^{-18}$ s$^{3/2}$); however, the high absorption coefficient limits application of this material due to insufficient optical quality of crystals.

After Abrams and Pinnow investigated acoustic-optical properties of germanium in 1970, it became the most popular material for modulation of IR radiation owing to the high optical quality, high value M_2 (840×10^{-18} s$^{3/2}$), high heat conductivity, and low acoustic losses.[202] Germanium's drawback is in enhancement of optical losses caused by absorption on free carriers at crystals warming up over 25°C.

The first evidence for ChG as the new category of material for acoustic and acoustic-optical application appeared in 1970 in the work by Krause and others.[203] In 1972 the works by Ohmachi and Ichida appeared, who investigated the properties of vitreous As_2Se_3.[204] Warner and Pinnow compared modulators on vitreous As_2Se_3, $Ge_{33}As_{12}Se_{55}$, and the crystal GaAs on the radiation wavelength 1.05 μm and showed that they all meet the requirements for pulse modulations.[205] Acoustic characteristics of some ChG of elementary systems were studied by Sheloput and Glushkov.[206] In 1973 to 1974 patents of the U.S. and the U.K. appeared, where it was recommended that glasses based on phosphorous, arsenic, germanium, tin, sulfur, and selenium be used as acoustic-optical media. Krause and others investigated the acoustic-optical properties of commercial ChG, manufactured by the company "Amorphous Materials" (U.S.), namely, $Ge_{33}As_{12}Se_{55}$ (TI20) and $Sb_{12}Ge_{28}Se_{60}$ (TI1173). According to their data the glass TI20 has $M_2 = 164 \times 10^{-18}$ s$^{3/2}$; the propagation rate of longitudinal acoustic wave is equal to 2.51×10^5 cm/s for TI20 and 2.38×10^5 cm/s for TI 1173. Ultrasound losses in TI 20 are 7.1 dB/cm.

In 1973 Ajio and Kislitskaya, in cooperation with Asnis and Andrianova, began to study acoustic-optical properties of ChG. Acoustic-optical characteristics were investigated, along with how they change with the composition of glasses in the systems As–Se. As–Ge–Se, As–Sb–Ge–Se, and As–Ge–Se–Te, and also the influence of thallium addings to glasses containing Ge–Se and Sb–Ge–Se. The data were collected of acoustic-optical properties of commercial glasses IRG29 and IRG34. Ultrasound wave rates, their damping, and coefficient of acoustic quality M_2 were investigated.

Determinations of ultrasound rate (V) were performed by the acoustic-optical method by the appearance of a diffraction pattern of stationary waves and the measuring of frequency interval between adjacent maximums (Δf). Attenuation factor of ultrasound (β) was measured by the pulse method as the intensity of diffracted light at acoustic wave propagation for a distance of 10 mm decreased. The measurements had been performed on the primary frequency of ultrasound radiator 18.2 MHz and on frequencies appropriate to the third, fifth, and seventh harmonics.

To determine M_2 the Dixon technique was used,[207] but the measurements were performed in Raman-Nat diffraction regime. It allowed one to work at lower frequencies, where ultrasound damping is low, and also to alleviate requirements to samples' homogeneity. In all cases the measurements were made on the elements for which there could be stated a clear diffraction pattern of ultrasound stationary waves at a frequency of 20 MHz.

Ultrasound rate was measured accurate to 1%. Coefficient M_2 and ultrasound attenuation factor were measured to 2.5% accuracy. Due to optical heterogeneity of samples, the variability of figures for the glasses of one and the same composition were 5% at the measuring of ultrasound rate and 10 to 15% for the determination of acoustic-optical activity and damping of ultrasound amplitude. The parameters noted above were calculated by the following formulas:

1. Ultrasound speed $V = 2l \times \Delta f$, where
    l is length of a sample following the direction of ultrasound propagation;
    Δf is frequency interval, in which the diffraction pattern appears.
2. Average photoelastic constant $Pav = 0.35 (1-\Lambda)$, where
    $\Lambda = -\rho/\alpha \times d\alpha/d\rho$, α is polarizability.
3. Average value of the coefficient of acoustic-optical quality $M_2 = \dfrac{n^6 Pav}{\rho v^3}$,
    where n is refractive index of glass; p_{av} is average photoelastic constant;
    ρ is glass density; V is speed of ultrasound wave's propagation in glass.

Mainly vitreous selenium As_2Se_3 is studied in the system As–Se, but this compound crystallizes in the regions of homogenization and cooling temperatures, thus limiting broad commercial application. On the contrary, other glasses of this system are quite stable, as well as glasses of the As–Ge–Se system.

In the system As–Se we studied acoustic-optical properties of glasses with content of As equal to 10, 20, 30, and 40 at.%; in the system As–Ge–Se — glasses with the constant content of arsenic equal to 20 at.% in which there had consecutively been an increase in the amount of germanium inserted at the expense of selenium. The influence of arsenic replacement with antimony proceeded based on the glass $As_{20}Ge_{12.5}Se_{67.5}$, where arsenic had been consequently replaced by antimony up to the composition $Sb_{20}Ge_{12.5}Se_{67.5}$. Recall that in complex systems, as distinct from the binary As–Se, physicochemical properties don't change monotonically: on the "composition-property" curves fractures are observed which confirm the accumulation of qualitatively new chemical bonds and new structural units in appearance.

The same nonmonotonous changes are observed on acoustic-optical properties as the fractures in the same compositions appear. As is seen from Figure 59, ultrasound speed (V) in the system As–Ge–Se increases as the content of germanium enhances. It is apparently connected with the decrease of an average atomic weight and increase of covalent bonds' number in a volume unit. Ultrasound damping (β) reduces therewith. Acoustic-optical activity also decreases, which is mainly connected with the refractive index decrease.

For arsenic replacement with antimony (Figure 60), an abrupt increase of density and refractive index of glasses occurs, as the result of a heavy element insertion which is followed by considerable gain in acoustic-optical activity. Ultrasound speed also slightly increases due to density enhancement. By the measured values $M_2\|$ and $M_2\perp$, where $M_2\|$ is coefficient of acoustic-optical activity when an electric vector of electromagnetic wave E is parallel with the direction of ultrasound propagation and $M_2\perp$ is perpendicular with it, elastooptic coefficients p_{11} and p_{12} were calculated.

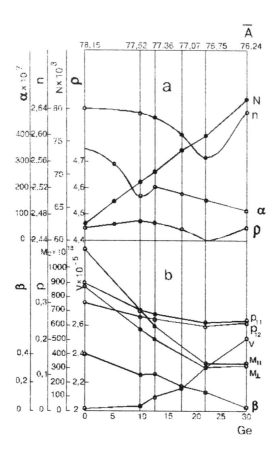

FIGURE 59. Dependency of acoustic-optical properties of the system $As_{20}Ge_xSe_{80-x}$ on Ge content.

With the increase of an average atomic weight ($\overline{A}$), $M_2\|$-$M_2\perp$ decreases, and hence p_{11}-p_{12} is enhanced.

In the system As–Se, when the content of arsenic increases, hence the density also does; ultrasound speed $V \times 10^{-5}$ cm/s rises from 1.85 to 2.2. Damping, accordingly, decreases. It shows up most vividly at high frequencies.

Acoustic-optical activity, if compared to As_2Se_3 in glasses with a lower content of arsenic, increases by 25% maximum. All these glasses have a value M_2 not less than 880×10^{-18} s$^{3/2}$ (i.e., approximately equal to germanium M_2), hence they can be used in acoustic-optical instruments. For pure antimonic glass $Sb_{20}Ge_{12.5}Se_{67.5}M_2 = 900 \times 10^{-18}$ s$^{3/2}$. At the same time the increase of germanium content leads to the considerable rise of ultrasound speed which determines fast operation.

Based on the obtained results, we come to the conclusion that glasses with the maximum content of germanium and antimony are of most interest for the simultaneous increase of effectiveness and fast operation.

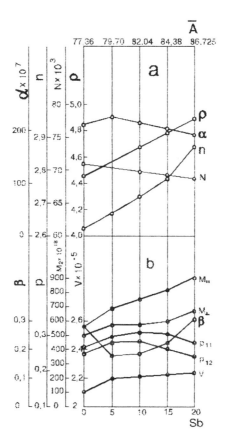

FIGURE 60. Dependency of acoustic-optical properties of the system $As_{20-x}Sb_xGe_{12.5}Se_{67.5}$ on the content of Sb.

Thallium introduction into the binary system As–Se causes an abrupt increase of M_2 up to the values exceeding 5000×10^{-18} $s^{3/2}$ in glasses with the constant content of arsenic equal to 30 at.% at consequent enhancement of thallium content up to 20 at.%. It can be explained not only by the higher refractive index of thallium glasses and low ultrasound speed, but also by the proximity of a wavelength of measurings (1.15 μm) to the absorption band.

Acoustic-optical properties of the glass $Ge_{22.5}Tl_{25}Se_{52.5}$ were investigated in the system Ge–Se–Tl. In the system Sb–Ge–Se–Tl as the initial composition $Sb_{10}Ge_{15}Se_{75}$ had been chosen, in which selenium was consequently replaced by thallium. The choice of compositions for the investigation was determined by the fact that up to 25% Tl all the replacement compositions are located in the center of the glass-forming region, which is important at the synthesis of sufficient amounts of glass of the required optical homogeneity. To improve reliability of measurements two samples of each composition were made which required not less than 40 g of glass. The synthesis regime, allowing one to produce samples of the desired homogeneity, was determined after viscosity and crystallizability of the glasses of chosen composition had been studied. Quality of the elements for acoustic-optical properties measuring was controlled with the help of examination under the microscope and spectral characteristics.

The change of ultrasound speed and coefficient of acoustic-optical quality coming from thallium content is shown in Table 27. Here the values of refractive index and density are also given. As is seen from the table, speed of longitudinal acoustic waves changes only slightly after thallium insertion into the glass $Sb_{10}Ge_{15}Se_{75}$: within 2.2 to 2.3×10^5 cm/s.

TABLE 27
Ultrasound Speed (V) and the Coefficient of the AO Quality of Thallium Glasses

Composition (at.%)				$V \times 10^{-5}$	$M \times 10^{18}$		ρ
Sb	Ge	Se	Tl	(cm/s)	$(s^{3/2})$	$n_{10.6}$	(g/cm^3)
10	15	75	—	2.2	490	2.551	4.62
10	15	70	5	2.3	590	2.667	5.07
10	15	65	10	2.28	630	2.784	5.45
10	15	60	15	2.24	850	2.9	5.81
10	15	55	20	2.23	1000 ÷ 1050	3.017	6.07
—	22.5	52.5	25	2.1	~1600	—	6.11

An average photoelastic constant p_{av}, necessary for the calculation of the coefficient of optical quality, was calculated by the above-noted formula. It is known, that for ChG of a broad composition range, where the atomic concentration changes within 0.055 to 0.059, the value Λ is practically constant and is equal to 0.15. Hence p_{av} for all investigated glasses is equal to 0.297. The values V and p_{av} don't change after thallium insertion; the principal contribution to the value M_2 is made by the refractive index. Its increase determines the M_2 rise from 490 to 1050×10^{-18} $s^{3/2}$.

As mentioned above, for modulating devices it is necessary that the medium possesses both high speed of elastic waves and high acoustic-optical activity. The investigations have shown that of most interest for the simultaneous increase of activity and fast operation are glasses with the maximum content of thallium, germanium, and antimony, because the increase of germanium content leads to the considerable rise of ultrasound speed, and antimony and thallium enhance acoustic-optical activity, changing only slightly ultrasound speed. In this way, thallium-containing highly effective glasses can be used for radiation control in low-powered lasers.

To investigate AO properties of commercial glasses samples from meltings of the first categories of quality were prepared. The properties of glasses are given in Table 28. It is seen that the values M_2 of glasses are rather high, both in the near IR-spectrum region and at the wavelength 10.6 μm; therewith in the glass IRG29 they are slightly higher than in IRG34. On the contrary, acoustic speed of the glass IRG29 is lower. Acoustic losses in IRG34 are the least. In this way the glasses

IRG29 and IRG34 may be recommended for manufacturing of acoustic-light guides for acoustic-optical deflectors. To provide high acoustic effectiveness IRG29 is preferable; as to fast-operating systems and also the devices operating at frequencies up to hundreds of megahertz, it's better to use IRG34 here.

TABLE 28
Acoustic-Optical Properties of the Glasses
IRG29 and IRG34

Properties of glasses	IRG29	IRG34
Speed of linear ultrasound waves $V \times 10^{-5}$ cm/s	2.18	2.45
Coefficient of the AO quality		
$M_2 \times 10^{18}$ s$^{3/2}$, $M_2(\perp)\ \lambda_{1.15}$ μm	291–318	311
$M_2(\|)\ \lambda_{1.15}$ μm	408–445	350
Ratio $M_2(\perp)/M_2(\|)$	1.405	1.12
Photoelastic constant P_{11}	0.157–0.168	0.161
Photoelastic constant P_{12}	0.135	0.153
Ratio P_{11}/P_{12}	1.183	1.052
Coefficient of the AO quality $M_2 \times 10^{18}$ s$^{3/2}$ M $(\|)\ \lambda_{10.6}$ μm	329–358	286
Coefficient of the attenuation on 55 MHz, cm^{-1}	0.314	0.039
Density ρ (g/cm^3)	4.745	4.475
Refractive index $\quad n_{1.15}$	2.685	2.675
$n_{10.6}$	2.595	2.589

4   Technological Basics for Manufacturing Optical Chalcogenide Glasses

4.1   INTRODUCTION

Before describing the peculiarities of ChG technology, it's necessary to understand how optical glass making differs from glass manufacturing for other multiple applications (construction, electrovacuum, glassware, etc.). The technological process for optical glass manufacturing is dictated by special requirements imposed on their properties and quality. The principal property of the optical glass, which sets it apart from glasses of other applications, is its high homogeneity. Optical glass must have a homogeneous composition, i.e., it shouldn't contain inclusions (threadlike or laminated) with a refractive index different from that of the major glass melt, so-called striaes. There must also not be smooth chemical heterogeneities, which cause continuous changes of the refraction index in the glass volume. It must also be free of heterogeneities arising from irregularities of a physical character, namely, double refraction, caused by mechanical stresses, and structural heterogeneities. The conjunction of all three above-mentioned factors — striaes of various origin, double refraction, and structural heterogeneity — determines the optical homogeneity of glass, i.e., equal absolute value of the refraction index and dispersion not only in the glass of the given melt, but also in half-finished articles of various batches. Apart from high homogeneity optical glass must have a maximum transparency in the effective wavelength interval. All these parameters are among the standardized quality indices of optical glass and they determine the necessary level of its technology.

The standard technological process of optical glass manufacturing includes the following stages: (1) equipment preparation, (2) loading and digesting of raw material mixture (of charge), (3) melt clarification, (4) cooling, (5) finishing. The time-temperature schedule at all stages is established for each glass, respectively. At the stage of loading the charge and digesting it is very important for the physicochemical processes, occurring during heating, to be thoroughly completed. Of these we refer to moisture and physically implicated gas removal, solid particles dissolving in the melt, polymorphous transformations, solid-phase reactions and reactions in the melt, etc. Naturally, complete charge melting, as a rule, is still in progress at the next stage, that is, clarification.

Clarification is the process of removal of bubbles from the melt, and glass melt composition alignment. During clarification the temperature in the furnace is raised to the maximum and then the mixing begins. The optical glass melting process, apart from glasses for other applications, is certain to include the stage of compulsory glass melt homogenizing through mixing by special stirrers to liquidate not only bubbles, but striaes as well. It is also an aid to the acceleration of glass-forming reactions. To observe the clarification process, every 0.5 to 1 h samples are taken out near the surface layer of the glass melt. The clarification process is recognized as complete when bubbles in a sample are reduced to the standardized number. After that comes the next stage of the technological process, namely, glass melt cooling. This stage is very important, for the optical homogeneity of glass and to a large extent depends on correct cooling conditions.

At the cooling stage mixing is still in process. The stirrer speed therewith reduces as viscosity increases, and mixing ends when the temperature matches the viscosity of about 500 to 100 P. Then the stirrer is quickly removed and the pot is taken out of the furnace at the melting temperature of about 900 to 1100°C.

The final stage — glass melt finishing — can be performed in two ways: cooling in the pot or melt casting into the mold or onto the table with rolling at the plate. After that the glass is put through the so-called "rough" annealing, i.e., slow cooling for relief of thermal stresses, leading to congealed glass breakage.

The final stage of optical glass manufacturing is "fine" annealing, which improves optical glass homogeneity in every respect.

In the late 1950s Danyushevsky worked out the mathematical theory of glass linear annealing.[208] The annealing is called linear when the speed of essential cooling is constant. When fine annealing is performed correctly, all continuous and structural heterogeneities are removed. The annealed glass in half-finished articles is put through quality control according to standardized parameters. Physicochemical basics of optical glass manufacturing are taken up in detail in the collection edited by Dyomkina, published in 1976.[209] All the above must be taken into account when organizing the manufacture of chalcogenide optical glasses.

4.2 MANUFACTURING TECHNOLOGY FOR COMMERCIAL OPTICAL ChG, PRODUCED JOINT-STOCK COMPANY "AOLZOS"

The production technology of chalcogenide glasses differs from that of oxide ones mainly due to the necessity of removing from glass melt traces of oxygen and hydrogen, present as chemical compounds, namely, oxides of various elements (extrinsic and intrinsic), hydroxides containing OH-groups, sorbed molecular moisture, etc. Also intolerable are impurities of carbon and carbon compounds.

Together with engineers Yakovlev and Geychenko, we elaborated on the industrial technological process, based on the laboratory synthesis technique (Chapter 1). In this way, chalcogenide glasses are industrially manufactured in evacuated vessels with the glass melt stirred by glass furnace rocking. That predetermines the series of changes in the process of melting in comparison with oxide glasses; in particular, it precludes observing the glass melt and glass quality control in the process of melting. A plant for chalcogenide glass manufacturing, according to the technology requirements, consists of three sections: (1) raw materials purification and charge preparation, (2) evacuation and melting, and (3) molling and annealing. The stages of the technological process don't differ from those for oxide glasses. Let us consider each of them, separately.

Purification of raw materials and charge preparation — At charge preparation the application of raw materials is provided, containing not less than 99.99 mass% of the major substance. Substandard raw materials are put through the repeated treatment. Vitreous granulated selenium (less common in plates) is derived from the manufacturer, sufficiently pure from metal impurities, but it may contain considerable amounts of oxygen and hydrogen. That's why each batch of selenium is tested for spectral characteristics. The useful transmission for glass-making selenium must not be less than 60% in the 3- to 15-μm range at 3 to 4 mm of sample thickness, with the minimum in the absorption band 13.5 μm not less than 30%. Selenium with lower transmission or with the presence of extrinsic absorption bands is put through preliminary purification by distillation in the regime of dynamic evacuation. Tellurium must be purified by the method we have elaborated. Metallic arsenic also contains, as a rule, oxygen compounds, because the control by these parameters isn't provided by the manufacturers. Hence arsenic purification is necessary. The purification process involves two stages: the cleaning of low volatile and, partly, of highly volatile impurities; and the final cleaning of highly volatile impurities. Arsenic is purified through sublimation in the previously evacuated unsoldered ampules which are broken just before charge preparation. Other components aren't supplementarily purified. As composition correction during the process is impossible, ChG making, the glass makers' responsibility increases for the accuracy at charge preparation, suspension, and filling. These works are mechanized as much as possible at plants.

Evacuation and glass-melting — Glass-making vessels are made of transparent quartz in the form of cylindrical bulbs with narrow necks and bouvvelet for connection with a vacuum-pumping assembly. To melt 4 kg of glass vessels Ø100 and 350 mm long are used; as the amount of simultaneously melted glass increases, the dimensions of vessels, naturally, enlarge. Glass-melting vessels are thoroughly cleaned of impurities by washing with hydrofluoric acid, water, and alcohol; after that they are baked out under vacuum (t = 800 to 850°C) to remove traces of moisture and oxygen. The prepared vessels are charged, connected to the vacuum-pumping assembly, and evacuated first at room temperature; then the charge is warmed up by the special regime. Warming the charge in laboratory conditions was introduced to remove hydrogen sulfide and hydrogen selenide and also SO_2, SeO_2, and other gases, forming in the process of digesting the charge. After warming the ampul is unsoldered from the vacuum-pumping assembly with the help of a gas burner and placed into the glass-making furnace. Charging and unloading of a glass-making assembly is mechanized. The glass-making furnace represents the tube electric device with the tube diameter appropriate to the dimensions of the glass-making assembly. Glass melt stirring is performed by furnace rotation at the rate established for each glass, respectively. The time-temperature regime of melting is also worked out for each glass individually; however, the principal stages — typical of optical glass making, namely, charge digesting, melt clarification, glass melt cooling, and finishing — are completely preserved. As was mentioned above, the charge digesting already begins during the preliminary warming up and is continuous for periods of time from several hours to several dozen hours, depending on the amount of glass and its composition. Here we would like to note again the necessity of technologically effective glass compositions for mass production. If some substance or blend can be synthesized in vitreous state in the amount of 10 or even 100 g, it doesn't mean that at the most correct technological process it can be produced as technical, or more so as optical material. The glass of technologically effective composition must be easily melted, have low volatility, and, most important, mustn't crystallize in the temperature range appropriate to viscosities necessary for melting and cooling. Certainly, an incorrect technological process can spoil any glass, but it is expedient to work out the time-temperature melting regime of industrial manufacturing only for technologically effective compositions. Figure 61 shows the standard diagram of melting regime and rough annealing of ChG as is usually designed in process charts and operating diagrams. Here Figure 61a shows the time-temperature regime of melting — stage I: charge complete melting; stage II: glass melt clarification; stage III: the temperature decreases to an annealing one. The regime of mixing is marked with broken lines. Figure 61b shows the time-temperature regime of rough annealing. When the melting is over, the glass-making assembly is taken out of the furnace and the glass is put through the final stage finishing. However, prior to that the quality of the glass is tested for imperfections and spectral transparency. For this purpose a piece of glass sufficient for making a control sample 8 to 10 mm thick is separated from the glass ignot taken out of the melting vessel. The sample is put through the cooling treatment and then tested. Under the microscope the number of insertions on 1 cm^3 is calculated and their dimensions are measured, then the spectral characteristic is taken and compared with the standard one to evaluate the categories of the glass of the given melt by IR transparency. The ready-to-use glass may be sliced into half-prepared articles by two methods: the cool one (ignot cutting into the half-prepared articles of the necessary size); and the hot one (heating up in the mold filled with glass melt either by gravity, so-called molling, or with premolding). The hot method is more efficient, because glass losses are reduced considerably. Immediately after molling the bars are put through fine annealing.

Molling and fine annealing — The following demands are imposed on the ChG molling and annealing furnaces: Heating up of the furnace with loading up to 600°C mustn't exceed more than 4 h, and furnace cooling from the maximum temperature down to 300°C must proceed as fast as possible. The glass is held in the temperature regime of crystallization for not more than 2 to 3 h, which prevents crystals in bars of commercial glasses. In molling furnaces, assembled at plants, the articles can be manufactured up to 500 mm across. The maximum diameter of bars produced nowadays is equal to 350 mm at a thickness of 45 to 50 mm and is determined by the amount of simultaneously manufactured glass. Prior to molling the glass ignots are put through an operation called dismantling. The ignots are thoroughly examined and swept free of surface defects. After that the glass is thoroughly washed with water and alcohol. An ignot, ready for molling, mustn't have cracks and deep cavities

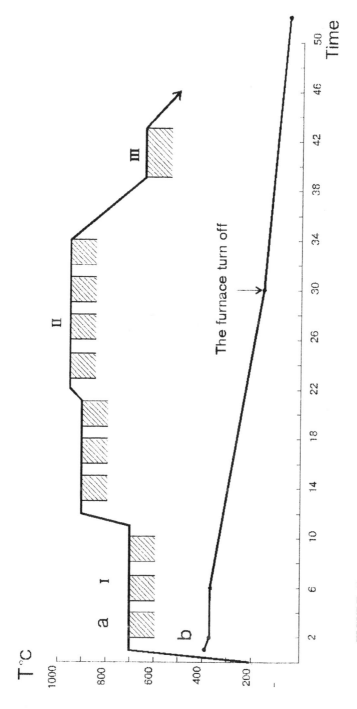

FIGURE 61. Example of the standard melting regime and rough annealing of commercial glasses of the IRG type.

as they cause inadmissible defects in bars. The industrial technology of flat detail molling doesn't essentially differ from the one worked out in the laboratory. The molling of bars of hemispherical shape is performed in two steps. The first molling by gravity gives to ignots a form close to that of a hemisphere. An ignot therewith is heated up to the temperature overcoming that of liquidus, hence crystal nuclei, sometimes formed at glass-making, are removed. The glass produced as a segment after the first molling is visually controlled. Defects are removed and glass weight is finally mated. The second molling is carried out with premolding. For this purpose the glass segment is put into a mold made of infusible material, which provides a bar of necessary size, around which metallic equipment is installed. That presents favorable conditions for a temperature equalizing by a bar. The metallic punch of a required size is connected with guides and heated. After heating up the furnace with a bar and heat time according to the molling regime appropriate to glass standards, the dome is opened, the equipment removed, and the heated punch is put into the mold and pressed in glass up to the stop, fixing it in this position by additional load. After the mold with glass is cooled down to a temperature 70 to 80°C higher than that of annealing, the punch is taken out and the mold in equipment is covered with the furnace dome and fine annealing is performed according to the regime. Figures 62 and 63 show the reproducibility of the spectral characteristics of the glasses of commercial meltings dated back to 1966 to 1967, of the first and second quality categories by IR-transparency. As seen from the figures, in those years the plant were unable to produce the glass IRG25 without absorption bands, and the reproducibility of the spectral characteristics was obviously insufficient. Such glasses were mainly used as inlet windows, protecting hoods, filters, etc., where the optical homogeneity wasn't standardized. In the 1970s, when the application began of ChG as the details forming a picture, the requirements on optical homogeneity began to be imposed on them. That, in turn, called for elaboration of new devices and methods which could allow one to evaluate homogeneity of details of various sizes made out of low transparent materials, or of those which don't transmit radiations in the visible spectrum range at all.

4.2.1 INVESTIGATION OF STRIAES IN CHG AND TECHNOLOGICAL WAYS OF THEIR LIQUIDATION

The so-called striaes are heterogeneities of refractive index, which appear in an article due to local changes in composition and represent one of the most dangerous defects of optical glass which sometimes absolutely distort the picture. That is why the striae control is obligatory in manufacturing commercial glass. For glasses, transparent in a visible spectrum, shadow and interference methods exist for detecting striaes. The shadow method is the easiest one — the so-called "method of a shining point". In this method the tested glass is placed in the cone of rays, emanating from the point light source and on the screen situated behind the sample, and the shadow projection of striaes is observed. The evaluation of striaeness of commercial ChG was performed from the photographs of striaes, with the help of the IR-shadow apparatus, specially elaborated by Doladugina and others, and calibration cords for $\lambda = 1$ μm. The description of equipment, measuring technique, and method for

τ %

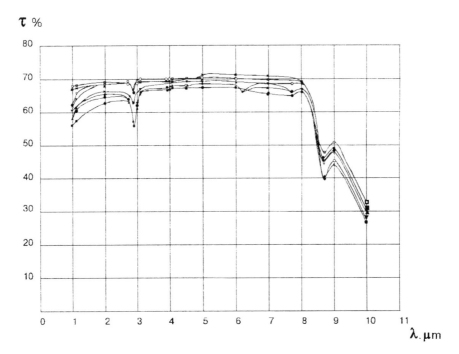

FIGURE 62. Reproducibility of spectral characteristics of the glass IRG23 from commercial meltings of 1966 to 1967.

the calibration of striae preparation is given in the research.[210,211] For the investigation six glass brands were chosen, which are in the most common use in IR optics: IRG23, 24, 25, 28, 29, and 30. The investigations were carried out on bars of series production shaped as disks or rectangular plates from 60 to 240 mm in size. The striaeness was tested in both bars of finished production; then the bars were put through single-stage molling and in the plates of the intermediate stage of finishing, which, for this purpose, were grinded and polished. For 2 years about 100 meltings were investigated, on the average 15 to 20 meltings of each glass brand. To evaluate the striaeness glass samples were also used, obtained 5 to 10 years ago. The glasses had been manufactured by normalized regimes, where control was performed by route charts.

Analysis of the quality of the glass IRG23 demonstrated that approximately 90% doesn't have striaes, hence the normalized regime of this glass manufacturing may be considered as close to the optimal one. On samples of IRG23 the influence of molling on glass quality as it concerns striaes was studied, and it was stated that in the molling process they are not formed. Unlike IRG23 all the samples of glass IRG24 investigated appeared to be striaed, therewith as a rule flows of striaes of one to two categories pass through them, and their refractive index is lower than in the major glass melt.

In individual cases laminated flows of rough striaes were observed, with refractive index being higher than that of the glass. The presence of striaes of various

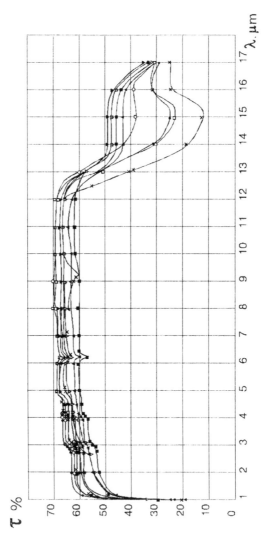

FIGURE 63. Reproducibility of spectral characteristics of the glass IRG25 from commercial meltings of 1966 to 1967.

character was also revealed in other glasses. The results of the analysis of the investigated glasses' quality are given in Table 29. The data demonstrated that at least four of the most commonly used ChG types need reconsideration of time-temperature melting regimes, for the glasses produced at the plant were not, in fact, optical and couldn't be applied in precision devices. This situation existed up to 1973 when Ajio, in collaboration with production engineers of the plant, completed work on the improvement of melting regimes for all commercial glasses. It was based on the investigation of viscosity at glass melt homogenization. The value of viscosity, corresponding to homogenization temperatures, was obtained through the extrapolation of the results of measurements on the impression viscometer. It was stated that the last mixing for glasses of various types takes place at very different viscosity values.

TABLE 29
Striae in Commercial Glasses

Glass type	No. of investigated meltings	Quality of glasses; type of defects	Absence of striae (%)	Viscosity at the last agitation (P)
IRG23	20	Unitary striae of the second category $n_{striae} < n_{glass}$	~90	~75
IRG24	15	Flows of striae over the second category $n_{striae} < n_{glass}$	0	~100
IRG25	15	Nodal striae, striae flows over the second category $n_{striae} > n_{glass}$	~10–15	~5
IRG28	15	Striae type is similar to IRG25	0	~5
IRG29	18	Unitary striae of the second category, threadlike opaque defects, oriented along the striae; $n_{striae} < n_{glass}$	~40–50	~50
IRG30	15	Central rough striae; $n_{striae} < n_{glass}$	90	~50

The glass IRG23 was obtained striae-free in 90% of the meltings and, besides, if compared to other glasses, it has the lowest crystallization tendency in the interval of synthesis temperatures; it was chosen for investigation of how the speed of mixing and the interval of admitted melt viscosity values influence nonstriaeness. These investigations were performed in the laboratory. Samples were made from ingots, weighing 120 g, of the glass melted simultaneously by the standardized regime with mixing for 30 min, but with different speed of the furnace rotation. It was stated that increase and decrease of mixing speed in comparison with the standardized one lead to the abrupt decline of glass quality by striaes and even to its stratification. The influence of glass viscosity on the process stage at mixing was found on the plates of intermediate conversion of commercial glasses — as a rule, $170 \times 170 \times 35$ mm, or on disks Ø150 mm, 30 to 35 mm thick. The mixing was being performed at several values of glass melt viscosity: poise lots, ~400 and ~1000 P. It is shown that the viscosity decrease down to poise lots leads to formation of many striaes rougher than the second category, whereas the viscosity increase up to 1000 P doesn't spoil the quality of IRG23. Therewith the total melting time may be slightly reduced. It appeared that the glass IRG24, which had been manufactured up to the mid 1970s, was always permeated by the flows of rough striaes. It's more complicated to melt this glass than IRG23, as its softening temperature is higher and it has some tendency to crystallization in the process of melting. According to the standardized technological process the last mixing took place at a temperature appropriate to viscosity of about 100 P. It's quite evident that at such regime the striaes connected with heterogeneities of the composition either can't be blended or reappear in the process of cooling due to convective flows. That's why the influence of melt viscosity on the quality of IRG24 at homogenization was studied. Experiments with the glass IRG23, confirmed by the investigation of IRG24 samples, manufactured in 1961 to 1962 in the form of a plate $150 \times 130 \times 20$ mm and a disk Ø100 mm, demonstrated that mixing at viscosities comprising the units poise and less only causes the formation of rougher striaes. A series of experimental manufacturings with glass melt mixing was carried out at viscosities of 400, 1,000, and 10,000 P. It should be noted here that all ChG are "short", i.e., viscosity changes rather rapidly with temperature change. In particular, for IRG24 the temperature interval, appropriate for viscosity change from 400 to 10,000 P, is equal to only 80°C. Investigations demonstrated that, taking into account the crystallizability of IRG24, the best regime for it is mixing at viscosity from 100 to 1000 P. Nonstriaeness of the glass IRG24, melted in this regime, reached 95%. Figure 64 presents photos of the shadow pattern of striaes of IRG24 samples, melted in the initial and changed regimes. The improvement of melting regimes for other glasses had been carried out by the same scheme, taking into account data obtained for IRG23 and IRG24 and also the peculiarities of compositions of the glasses investigated. The only problems were with the glass IRG28, for its high crystallizing tendency doesn't permit the last mixing at the optimal viscosity value; that's why nonstriaeness was enhanced only up to 60%. However, as this glass is recommended for application as the filter in the third atmospheric window (8 to 14 μm), fine striaes, observed on the wavelength of 1 μm, don't influence the quality of the picture. The new technological process guarantees nonstriaeness equal to 90 to 95% for the majority of commercial glasses; hence in this parameter ChG reached the level of optical glasses. The quality of glasses according to cords is guaranteed by the technological process, but at equipment replacement, change of regimes, etc., the striaeness can also change. The examination of the striaes' shadow pattern is thus compulsory.

4.2.2 Physical Heterogeneities in ChG

As was mentioned above, striaes are not the only defects that influence the optical homogeneity of glass. In glasses so-called "physical heterogeneities" may also take place, including heterogeneities of structure which appear due to irregular glass cooling in the annealing area and photoelastic heterogeneities, which appear under the influence of residual stresses. The last derive from the different cooling speed of middle and edge zones of bars.

To liquidate structural heterogeneities optical glasses are put through fine annealing. The problems of fine annealing of oxide optical glasses are scrutinized in the

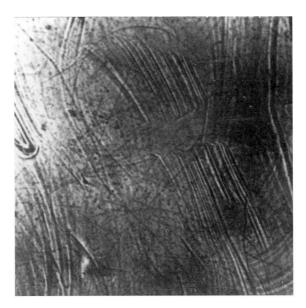

FIGURE 64a.

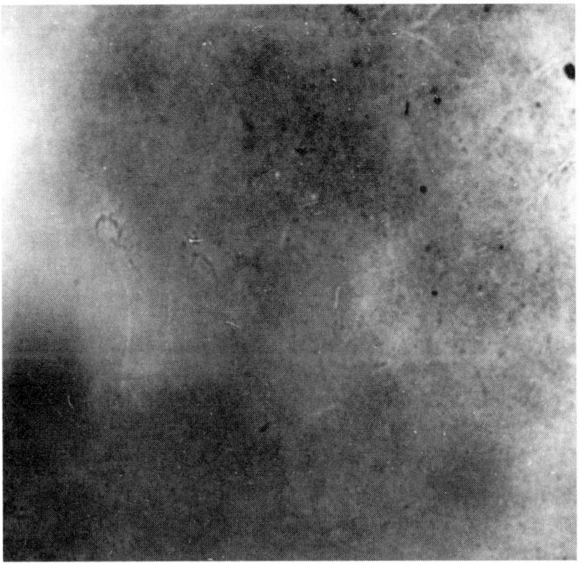

FIGURE 64b.

FIGURE 64. Shadow view of striae in the glass IRG24 before (a) and after (b) changes in the melting regime.

monography by Gomelsky.[212] According to the theory, at the upper annealing edge the temperature is taken, appropriate to the viscosity equal to 10^{13} P, at which 95% of all glass stresses have been relaxing for 2 min. The annealing area is the temperature interval where the relaxation of stresses takes place, i.e., where it's possible to create constant stresses or to liquidate the existing ones. For an average glass this interval is equal to ~150°C. Calculations have demonstrated that in this interval the viscosity changes by 5.5 orders of magnitude, hence the glass viscosity equal to $10^{18.5}$ P is appropriate to the lower annealing edge. According to Gomelsky's calculations, the temperature interval of the annealing area for glasses of various compositions varies from 60 to 160°C. In both annealing and finishing areas the glasses are divided into "short" and "long" ones. The refraction index n dependence of the cooling speed h is expressed by the Danyunevsky equation:

$$n = \frac{\beta}{K} \lg h + u \qquad (1)$$

where β is temperature coefficient of refractive index change. β_{abc} is defined as the refractive index change at temperature change by 1°C; K is constant, depending on the glass composition; u is constant, depending on glass properties. Having designated β/K as m, we have:

$$n = -m \lg h + u \qquad (2)$$

The coefficient m may be determined by the experiment if in samples of one melt-glass, cooled with different speeds h_1 and h_2, measure, respectively, n_1 and n_2. Then the originating difference in refractive indices Δn will be equal to:

$$\Delta n = n_1 - n_2 = -m \lg \frac{n_2}{n_1} \qquad (3)$$

This difference is called the relative annealing number. As is seen from the formula, the annealing number Δn, as well as the absolute value of the refractive index, depends only on the cooling speed in the annealing area and it doesn't depend on either temperature or holding duration, or size of bars. It was experimentally confirmed on examples of oxide glasses as long ago as in the 1950s.

In optical glass making at fine annealing six standard regimes of cooling are used with speeds 0.5, 1, 2.5, 5, 10, and 20°C/h. As the basic value the refractive index n_0 of glass cooled at a speed of 2.5°C/h is taken. Then $\Delta n - \lg h/2.5$ curve is plotted and, according to the theory, this dependence must be linear and with its help or by the table of "annealing numbers", based on the data of the curve, the change of the refractive index of the given glass type at annealing is determined. The value n after annealing is $n = n_0 + \Delta n$, where n_0 is the certificate value of the refraction value and Δn is the annealing number. The value Δn may be both positive and negative depending on whether the cooling speed h, appropriate to n, will be more or less than 2.5°C/h. In this way we can change the refraction index to necessary values in certain limits, varying the cooling speed.

The given theoretical assumptions hadn't been extended to ChG up to the mid 1970s. Annealing had been performed only to eliminate ultimate tensile stresses, which was enough for their application as windows and filters. In 1961, at normalization of the technological process of ChG manufacturing, according to the results of laboratory investigations, at the upper annealing edge temperature was taken appropriate to the viscosity $10^{11.5\pm0.5}$ P_a, which takes place between T_g and dilatometric softening temperature T_{ds}. The annealing shift into the region of viscosities which provides much more rapid structure relaxation allows it to perform cooling at a higher speed, which is especially important for "short" ChG. The cooling was carried out at a speed of 10°C/h. In 1975 Kanchiev made the first effort to determine experimentally the lower annealing limit.[213] By the example of ChG the author demonstrated that the first fracture on the dilatometric curves of the quenched fibers is appropriate to the viscosity $10^{18\pm0.5}$ P. At these very viscosity values, coinciding with the theoretical ones, the relaxation processes begin, connected with structure stabilization. The influence of the cooling speed for commercial ChG on the optical constants in the region of critical annealing had been studied on the glasses which were used for the chemical homogeneity investigation, and also on IRG31 and IRG32. Melting, molling, and annealing were performed at plants on industrial equipment. Glasses of the highest quality were mollated into bars for prisms, and after that cooled by six regimes at speeds of 0.5, 1.0, 2.5, 5, 10, and 20°C/h down to the lower annealing temperature. This temperature, appropriate to the viscosity $10^{18.5}$ P, had been found through the linear graphic extrapolation of experimental results of viscosity measurings in the interval 10^6 to 10^{13} P in coordinates gh − 1/T, and also using the data by Kanchiev. Out of the bars annealed at different speed, the prisms were made with large surfaces of 35 × 35 mm and refracting angle of 11°C ± 30′, on which the refractive indices were measured with the aid of the goniometric instrument, developed by Aijo. Reproducibility of the measurements of the value n = ±3 × 10⁻⁴. For all eight types of glasses it was possible to extend processes of linear annealing to ChG. For a standard the value n was taken at the wavelength 5 μm, where there is the lowest dispersion; the most stable graduation of monochromator prism and ambient temperature oscillations within 20 ± 3°C have no influence.

The experiment showed that the change of annealing regime for glasses of all types causes different variations of the refractive index, whereas the dispersion $n_{1.8}$ to $n_{2.2}$ and $n_{1.8}$ to $n_{11.0}$, in spite of the considerable difference in compositions, remains constant within the accuracy of measurements. That is why the results obtained will be considered by the example of the glass IRG29 (Table 30).

The dependence n = f(lgh) is linear for all glasses, i.e., the theoretical concepts of the linear annealing extends to ChG (Figure 65). At Δn calculation the refractive index of glasses annealed at a speed of 5°C/h was taken as the nominal one. Its value can be found through interpolation on the curve n-lgh (broken vertical line in Figure 65a). Then the graphical characteristic Δn = f(lgh/5) is plotted (Figure 65b), according to which the values of annealing numbers are found and the angular coefficient m is defined from Equation 3. The values of angular coefficients of ChG may vary over wide limits, unlike oxide glasses where m is, as a rule, (4 to 10)10⁻⁴.

TABLE 30
Values of Refractive Index and Peculiar Dispersions
of the Glass IRG29, Annealed at Different Cooling Speeds

Wavelength (λ μm)	n_λ at the cooling speed (°C/h)				
	0.8	1.2	1.7	4.4	5.5
1.8	2.643	2.6431	2.6426	2.6417	2.6406
2	2.6368	2.637	2.6364	2.6354	2.6346
2.2	2.6323	2.6324	2.6318	2.6309	2.6301
2.6	2.6264	2.6264	2.6259	2.6249	2.6242
3	2.6233	2.6233	2.6225	2.6215	2.6209
5	2.6138	2.6136	2.6129	2.612	2.6113
7	2.6087	2.6087	2.6077	2.6068	2.6062
9	2.6031	2.6029	2.6024	2.6013	2.6007
11	2.5968	2.5967	2.5959	2.595	2.5945
$n_{1.8}-n_{2.2}$	0.0107	0.0107	0.0108	0.0108	0.0107
$n_{1.8}-n_{11}$	0.0462	0.0464	0.0467	0.0467	0.0463

The angular coefficient of the glasses IRG32 and 34 lies within the limits 15×10^{-4} to 0 and in other glasses it reaches $(29 \text{ to } 52)10^{-4}$; it is evident that great structural changes take place in the annealing process (Table 31).

It is obvious that for ChG rate fixing by the refractive index they must be put through the essential (fine) annealing by the strictly followed linear regime, worked out in view of annealing numbers. At the same time large values of annealing numbers allow us to correct deviations of refraction indices, caused by composition rupture, selecting the annealing regime. To eliminate photoelastic heterogeneities it is necessary to know the residual stresses connected with the temperature drop. These stresses are calculated from the Adams and Williams equation, applied for oxide glasses by Gomelsky.[212] According to this equation the stresses P_x in the points of the plate of unlimited size, being offset by X to any direction from the middle surface of the plate, are connected with physicochemical properties of glass in the following way:

$$P_x = \pm \frac{\alpha E \eta}{6a(1-\mu)}\left(b^2 - 3x^2\right) \tag{4}$$

where b is half thickness of the plate in mm; α is coefficient of thermal linear expansion of glass; a is coefficient of thermal conductivity, cm²/min; E is elastic modulus, kg/cm²; μ is Poison coefficient; and h is speed of warming up and cooling,°C/min. To symbolize literal coefficient of the binominal $(b^2 - 3x^2) - \alpha E/a(1-\mu)$ as θ, we may write

$$Px = \pm \frac{\theta h}{6}\left(b^2 - 3x^2\right) \tag{5}$$

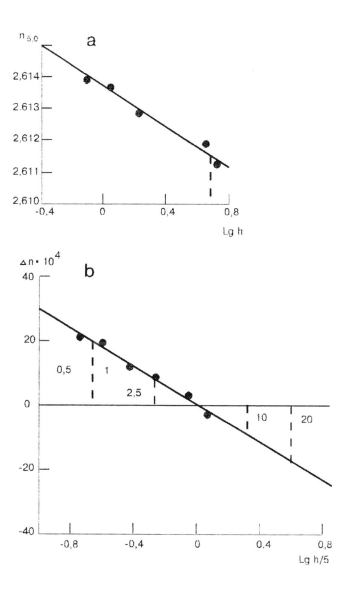

FIGURE 65. Dependence of the refractive index of glass IRG29 on the logarithm of the cooling speed h (a) and the diagram for annealing number determination (b).

Coefficient θ for each glass type has its own value. Coefficient θ/6 of industrial ChG and also the values, necessary for calculating it are given in Table 32. The calculation was performed up to the temperatures lying below the region of essential annealing, hence with no regard to the changes in glass property with temperature. In critical region, the properties greatly change there; that's why it's necessary to have in mind when calculating residual stresses that the relation α/a at high temperatures is, on the average, 2.2 times larger than at low ones. The double refraction may be calculated

TABLE 31
Changes of the Refractive Index (Δn_2) of Glasses Caused by the Speed of Temperature Decrease (Annealing Numbers)

Glass	$\Delta n \times 10^4$ at a cooling speed (°C/h)						Angular coefficient
type	0.5	1	2.5	5	10	20	(m × 10⁴)
IRG23	47	33	15	0	−14	−29	47
IRG24	52	37	17	0	−15	−31	52
IRG25	36	25	11	0	−10	−21	36
IRG28	36	25	11	0	−10	−21	36
IRG29	29	21	9	0	−8	−17	29
IRG32	15	10	4	0	−5	−10	15
IRG34	0	0	0	0	0	0	0

from the formula: $\delta n = BP_x$, where B is Bryuster coefficient. In this way, knowing the annealing speed and the thickness of the plate, we can define to sufficient accuracy the value of temporal and residual stresses. Normalization of chemical and physical heterogeneities is necessary to evaluate the optical glass quality and possibly their effective application. For example, while applying the glass IRG32 in the precision objective of the thermal vision device, instability of the line scattering function (LSF) was revealed. The lens made of this glass were investigated on the plant FM-115, which allows one to measure the dispersion in small angles (~1 to 2°) at wavelengths 1.8 to 3.0 μm. The investigation showed that the lens in different points has a different scattering coefficient which varies within the limits 10 to 22%. Therewith all certification characteristics of glass of given melt spectral transmission, optical constants, and attenuation rate in the region 8 to 12 μm met the first category. On the replicas, taken from the glass and photographed under the electron microscope at 14,000×, crystalline insertions 0.1 to 1 μm in the amount of approximately 1 to 2% were revealed. This fact made it necessary to investigate small-angle scattering in all commercial IRG. The samples 50 mm across of different thicknesses were made out of glass bars of commercial meltings. The polishing was performed on tar with accuracy of 3 N immediately after fine grinding. The samples were studied before and after one-ply clarification with maximum transmission at $\lambda =$ 2.5 μm. The results demonstrate that the glasses IRG23, 24, 25, 29, 30, and 34 have minimal scattering. It is quite clear that the glasses practically don't crystallize and the scattering in them is mainly defined by the number of dark insertions of different origin. In the glasses IRG28 and 32, which have a certain crystallization tendency in the process of synthesis, it's very difficult to achieve a scattering coefficient equal to 1. It is clear from the above that when manufacturing even slightly crystallizing glasses, strict adherence to time-temperature regimes is required, especially at the stages of molling and annealing. For the majority of the most commonly used glasses types the standardized technological process of melting and yield produces glasses with scattering not more than 3%, which is enough to keep LSF constant.

TABLE 32
Coefficient Θ/6 in the Equation for Residual Stresses Calculation

Glass type	Young's modulus (E × 10⁴ kg/cm²)	Coefficient of temperature conductivity (a × 10³ cm²/min)	Poisson coefficient (μ)	Brewster coefficient (B_nm.cm/kg)	Thermal coefficient of expansion (α × 10⁶ °C⁻¹)	Θ/6 = αE/6a(1 − μ)	BΘ/6 = BαE/6a(1 − μ)
IRG23	16	114	0.31	−10.2	24.6	8.14	−82.2
IRG24	19	120	0.28	−3.3	18.2	6.67	−19.8
IRG25	19	126	0.29	−31.6	22	7.79	−233.7
IRG28	18	126	0.29	−27	22	7.38	−184.5
IRG29	18	120	0.28	−22.4	22	7.91	−150.3
IRG32	27	99	0.26	−20.4	14.7	9.31	−186.2

4.2.3 INFLUENCE OF VARIATIONS IN GLASS COMPOSITION DURING THE MANUFACTURING PROCESS ON THE REFRACTIVE INDEX

The annealing numbers demonstrate how the refractive index changes when the annealing speed changes. However, if at control measurements considerable deviations of n are observed, the reason should be looked for in the instability of glass composition. Thus, in the application of glass IRG23 in the objective of one of the wholesale devices for thermal vision, in different batches of components deviations were found of the refractive index, which reached 150×10^{-4}, that destabilized the parameters of the device.

In sulfur-containing glasses we used trisulfide arsenic; the selective testing of the initial composition was performed. It appeared that in different batches of oripigment there were deviations from stoichiometry, which for arsenic reached 2 to 3 wt%. Earlier Pashinkin and others, while investigating the composition of vapor over the melt As_2S_3, revealed its essential dissociation on As_4S_4 and sulfur.[214] To find out how the instability of the composition As_2S_3 influences its properties, we synthesized the glasses of 12 compositions (including stoichiometric ones) in which the arsenic content varied from 57.5 to 64 wt%, i.e., excess and lack was +3.1 and –3.4 wt%. It was stated that at arsenic excess and lack the volume concentration of atoms N decreases, i.e., the structure of stoichiometric As_2S_3 appears to be the most consistently packed. In Figure 66 one can see how the refractive index of the glass IRG23 depends on the excess-lack of arsenic — using an example of the glass-analogues, melted at a plant in the appropriate regimes: even at deviations of 0.3 wt% the refractive index changes in ±50 to 60×10^{-4}, and it surpasses considerably the normalized changes. The calculation has shown that to produce glass IRG23 with the value n, variable within the allowable limits, deviations from the stoichiometric relation between arsenic and sulfur mustn't exceed ±0.05 wt%. That is why to produce sulfur-containing glasses with stable values of optical constants, the correction is performed at the charge preparation. The amount of excessive arsenic is defined by the ultrasound method, elaborated by Nemilov,[215] which presents the accuracy of definition up to ±0.03 wt%, and the proper amount of elementary sulfur is introduced into the charge. In selenium-containing glasses the refractive index may vary with excess or lack of selenium, though not so essentially. It is also seen from Figure 66 for example of glass IRG23. When the composition changes, the peculiar dispersions also change: an excess of arsenic leads to their enhancement; its lack — to their decrease. Selenium influence is the opposite. Everything mentioned above underlines once more that the most important factor at mass production of ChG is the stability of the composition, and its maintenance in different batches is especially significant when the amount of simultaneously melted glass is low.

4.3 TECHNOLOGY FOR THE PRODUCTION OF COMMERCIAL CHALCOGENIDE GLASSES MANUFACTURED BY DIFFERENT COMPANIES (REVIEW)

Not too many publications exist which describe in detail the ChG synthesis process even in laboratory conditions, as distinct from the enormous number of works where

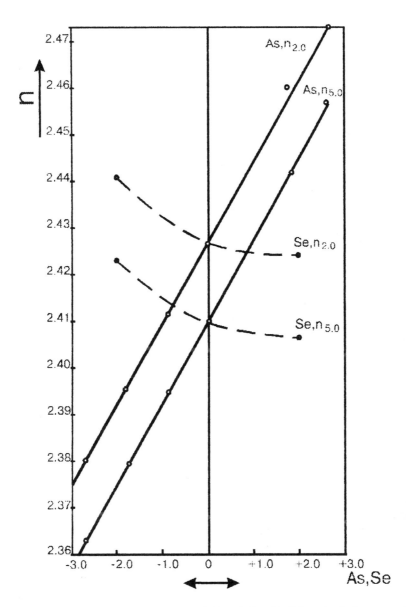

FIGURE 66. Dependence of the refractive index of glass IRG23 on the lack-excess of As and Se (wt%).

properties and structure are studied. As for the technological processes described for commercial ChG manufacturing, one can find only a few articles on this topic. After having analyzed the articles by Savage, Nielsen, and others we can come to the conclusion that the company Barr and Stround manufactures its glasses by way of vacuum synthesis from elements in silica tubes, at $T = 970°C$ in a rocking rotation furnace. After the synthesis is over, the glass is air-quenched, annealed, cut, and polished (see, for example, Reference 146). The authors don't present, however, any details of time-temperature regime, as well as of the glass-melt homogenizing. In the publications by Feltz and others a scheme is given for the manufacturing of the glasses, produced by the company Jenaer Glaswerke GmbH, using as an example the glass $Ge_{30}As_{20}Se_{50}$ with the mass equal to 1 kg.[216] According to this scheme the process begins with an ampul's purification with the help of $HF:HNO_3$ (1:1) and the subsequent warming up in vacuum 10^{-5} Torr at T = 900°C. Semiconductor germanium is weighed out into the prepared ampul. Selenium and arsenic are put through prepurification, because semiconductor granulated selenium contains not only SeO_2, but also a considerable amount of water; commercially available arsenic also isn't free of oxygen and moisture impurities.

The purification is carried out in the specially made double-walled ampul with a melted frit between the walls. Selenium is first warmed up to 300°C. As a result SeO_2 and water are distilled into the forepart, made as a sprout, which is annealed afterward. Then selenium is distilled through the frit into another part of an ampul at T = 500°C. The purified selenium is weighed and added into the melting ampul.

Arsenic is purified in the same way. Therewith an insignificant black deposit is always observed which doesn't dissolve in glass melt and causes light diffusion to insert nonsublimated arsenic. After being deoxidated and sublimated arsenic is weighed and inserted into the melting ampul under inert conditions. The ampul is evacuated to 10^{-5} Torr and annealed. The synthesis conditions, in particular, the time-temperature regime, are not presented here. The authors note that by this method compact, cylindrical, glass blocks are produced: they can be cut into disks up to 60 mm across.

The technological process for optical ChG, produced by the company Amorphous Materials Manufacturing Inc., is described in detail in the article by Hilton and Cronin.[181] This process, as well as some laboratory techniques, includes the purification of the ready glass by its distillation in vacuum, but it is more advantageous if the glass synthesis and distillation are performed in one installment without intermediate atmosphere contact. The installment consists of two chambers, connected by a tube containing a porous quartz filter. The reactants are loaded into the compounding chamber, which represents a long quartz tube, to which after the loading the cap is sealed, connected with the turbo molecular pump. Both chambers are evacuated, the reactants are heated to remove moisture, surface oxides, and other impurities from the melted selenium, after which the chamber is sealed. The casting chamber, prepared by sealing a polished quartz plate into a crucible, was 8 in in diameter, which enables it to obtain a 9-kg plate about 2.5 in thick. Nowadays the form and sizes of the casting chamber have, apparently, changed, for, judging by the catalogues, Amorphous Materials Company supplies the plates up to 12×18 in.

During the compounding process the casting chamber is kept at a high temperature to avoid premature glass transfer. Glass melt homogenizing during the compounding step is achieved by the rocking furnace rotation. After the compounding process the stirring is stopped, the temperature of the casting chamber falls down, and in consequence all the glass is distilled into the casting chamber through the quartz filter to remove particulate matter from the glass.

The rocking furnace is turned on again to ensure the condensed glass is properly mixed until a quenching temperature is reached. After the stirring is over, cold air is blown into the chambers to quench the glasses quickly to annealing temperature (370°C). Then the furnace is closed and held at this temperature for several hours. The whole process is controlled by a computer. After quenching the plates are checked for the presence of foreign particles and oxide absorption using an IR microscope and spectrometer. The acceptable plates are put through fine annealing. The annealed plates are ground flat and parallel for polishing, after that their transmission is measured in the wavelength region 8 to 12 μm. The optical path is 2.5 in, which makes the results very sensitive to absorption. A quality control sheet is provided to verify the quality of any glass blank. One can see that the glass quality control is performed in conformity with all necessary parameters, and the company has a right to guarantee the absence of oxides, foreign insertions, and striaes in the plates.

The properties and quality of commercial glasses, manufactured by different companies, are considered below.

4.4 PROPERTIES AND QUALITY OF CHALCOGENIDE GLASSES MANUFACTURED BY DIFFERENT COMPANIES

The data available to us about the commercial ChG are not, unfortunately, complete and reliable enough, for we haven't gotten all catalogues and folders of various companies.

We can imagine, however, the situation with the manufacturing of these materials, in general. Until the mid 1960s six companies had been manufacturing ChG for IR optics. They were Servo Corporation and American Optical Company, Barr and Stround (England), Schott and Gen (Mainz, Germany), which produced As_2S_3, Eastman Kodak (England) which produced the glass As_8Se_{92}, and Parra-Mantois (France), producing the glass As_2Se_5. The two last transmit radiations with wavelengths 11-12 μm, but their T_{ds} are too low (70 and 130°C, respectively). We don't know whether the glass As_8Se_{92} manufacturing survived.

In "The Buyer's Guide" — annual appendix to the magazine "Laser Focus World of 1994" — three IR glasses are advertised for laser optics, namely, As_2S_3, As_2Se_3, and KRS-5.[219] Since KRS-5 isn't glass but a crystal (44% TlBr, 56% TlI), let us cite the data about the glasses.

Both glasses are manufactured by six companies: Amorphous Materials Inc., Atomergic Chemetals Corporation, Newport Optic Materials Inc., Cerac Inc., Ispra Israel Product Research Company Ltd., Optical Materials Inc. Two more companies produce only As_2S_3 — Morton International La Plaine and Servo Corporation of America. We note that only the last company has been manufacturing this glass for more than 20 years. All the rest of the companies are comparatively new: as evidence from "The Buyer's Guide" (1977), 13 companies then manufactured As_2S_3, 11 companies As_2Se_3, but neither of them, except Servo Corporation, were among the manufacturers in 1994.

In the 1970s Texas Instruments began production of two glasses — TI 1173 ($Ge_{28}Sb_{12}Se_{60}$) and TI 20 ($Ge_{33}As_{12}Se_{55}$). Lloyd, in his monograph about the elaboration of the systems for thermal vision as optical IR materials, close to perfection, along with silicon, germanium, zinc sulfide, and zinc selenide, reports about the glass TI 1173.[217] Nowadays these glasses have new designations: AMTIR-3 (the same with TI 1173) and AMTIR-1 (TI 20). The glasses AMTIR differ from their analogues in extremely low values of the absorption coefficient, and also in slightly higher values of optical constants, which are determined by the new technology, published by Hilton and Cronin in 1986,[181] which was observed in detail above.

AMTIR-1, AMTIR-3, and highly pure As_2S_3 are produced now by the company Amorphous Materials (U.S.), which is, apparently, a filial of the company Texas Instruments. Their properties and qualities are described from the data of 1990 to 1991, based on the company's catalogues. We also have the catalogues of the company Barr and Stround (England), which advertises two glasses BS1 and BS2, and also As_2Se_3, with the designation BSA. The glass compositions are not given in the catalogues, but one can easily establish them using Savage's data: he presents optical constants, density, and CTE for a series of glasses of the systems Ge–As–Se, Ge–Sb–Se, and Ge–As–Se–Te including BS1, BS2, BSA, TI 20, and TI 1173.[105] The company Jenaer Glasswerke GmbH (Germany) advertises five glasses: IG 2n2919, IG 32930, IG 41921, IG-51922, and IG-61923. Optical and some physicochemical properties of glasses of the series IG were taken from Feltz and others' publication which appeared in 1991.[218]

The information about the properties of glasses IRG, produced by the Joint-Stock company "LZOS", are reported based on the catalogue of the State Optical Institute, published in 1990, and the standardized quality parameters according to the branch standard of 1983. The information about the glasses AV, IRG-35, and AST-1 were obtained from the investigations of experimental-commercial meltings of 1985 to 1990.

It is quite easy to compare optical properties of glasses, supplied by different companies, for the changes of spectral transmission are performed on spectrophotometers of the same or equivalent types by similar methods. The refractive index is measured in all cases by the goniometric method, with an accuracy of $\pm 1 \times 10^{-4}$; the particular dispersions are, as a rule, determined for two wavelength intervals $n_{3.0}$ to $n_{5.0}$ and $n_{8.0}$ to $n_{12.0}$; the dispersion coefficient is calculated from the formulas

$$V_4 = \frac{n_4 - 1}{n_3 - n_5}; \quad V_{10} = \frac{n_{10} - 1}{n_8 - n_{12}}$$

For the glasses of the State Optical Institute also added are the values of the peculiar dispersion $n_{1.8}$ to $n_{2.2}$, i.e.,

$$V_2 = \frac{n_2 - 1}{n_{1.8} - n_{2.2}}$$

The situation is more difficult with physicochemical properties. According to the literary data, only density is measured everywhere by the method of weighing the substance in toluene, hence, it is also comparable. The measurements of mechanical, thermal, and other properties are performed on different apparatuses, using different methods, and frequently the results are expressed in different values. This applies to even a greater extent for parameters of glass quality.

In consequence of the above, in the present monograph only the optical constants and transparency limits of all known commercial glasses are generalized in Table 33. For the glasses transparent to 14 μm, the Abbe diagram was also plotted in the coordinates $n_{10.0}$–$v_{10.0}$ (Figure 67). It is seen from the table and the figure that the glasses may be combined into several groups according to their optical constants. Thus, three groups of glasses have practically similar values n and v:

1. BS-1, AMTIR-1, IG2n (n = 2.4914, 2.4977, 2.4967; v = 113, 113, 105)
2. IG 5, AMTIR-3 (n = 2.6038, 2.6023; v = 102, 110)
3. IRG25, IG 6, BSA (n = 2.7675, 2.7775, 2.7758; v = 152, 161, 159)

Three pairs of glasses have similar values: IRG29, IRG34 (n = 2.6006, 2.5941; v = 122, 131, respectively); they are located in proximity of the glasses of group 2: BS2, AV2 (n = 2.8563, 2.8424; v = 185, 162); IRG32, AV1 (n = 2.9731, 2.9920; v = 113, 120). It should be noted, however, that glasses of the series AV are more technologically useful because they don't crystallize under manufacturing conditions. These glasses and the glasses IG4, IRG28, AV3, and AV4, which don't have analogues, both in combination with each other and with other glasses, yield large aperture crown-flint pairs, hence to correct all types of abberations, not using crystalline materials. The glass AV4 may be called IR superheavy crown, and the glasses AV-1 and IRG32 — superheavy flints.

In Table 33 the values are given of optical constants of the glasses As_2S_3 (Amorphous Materials), IRG23, and IRG24, transparent up to 9 to 11 μm and destined for application in the near- and middle-IR-spectrum range. These glasses transmit radiations in the far-red region of the visible spectrum range, which allows one to perform visual alignment of apparatuses. Their dispersion coefficient is rated at wavelength range $n_{1.8}$ to $n_{2.2}$ and $n_{3.0}$ to $n_{5.0}$.

The glass IRG35 was not included in the catalogue and is a custom order. It has the lowest refractive index ($n_{10.0}$ = 2.370) and rather low dispersion ($v_{10.0}$ = 274). The glass is used as an optical element for spectroscopy BAIR, but it can be applied as IR glue, immersion medium, and for production of optional optical details. Filters IRG27 and IRG33 occupy a particular place. The refractive index isn't so important for them, except for the cases when they are used as lens at the same time.

The values of the absolute temperature coefficient of the refractive index $\Delta n/\Delta T$ (β_{abs}) are presented in literature only for the glasses produced by the company Amorphous Materials Inc.; therewith for each of the three glasses different temperature intervals and wavelengths are given. We cite these data together with the values β_{abs} of glasses of the IRG series, which are middle ones in the temperature range

TABLE 33
The Optical Constants of Commercial Glasses

Glass type	Refraction index (λ μm)									Dispersion coefficient			Transparent range (μm)
	1.8	2	2.2	3	4	5	8	10	12	V_2	V_4	V_{10}	
IRG23	2.4303	2.4261	2.4232	2.4163	2.4124	2.4086	2.3965	—	—	201	183	—	0.7–9
IRG24	2.4134	2.4098	2.4062	2.399	2.395	2.3911	2.3806	—	—	196	176	—	0.8–11
TI20	—	—	—	—	—	—	2.5002	2.4942	2.4867	—	—	111	1–15
AMTIR1	2.5469[a]	2.531	2.525[b]	2.5184	2.5146	2.5112	2.5036	2.4977	2.4902	—	194	113	1–15
BS1	—	—	—	—	2.512	2.507	2.4972	2.4914	2.484	—	—	113	2–14
IG2n	—	—	—	2.5173	2.5129	2.5098	2.5024	2.4967	2.4882	—	202	105	0.8–12
IRG29	2.6443	2.6381	2.6333	2.6225	2.6183	2.6141	2.6065	2.6006	2.5934	149	193	122	1–15.5
IRG34	2.6339	2.6283	2.6241	2.6147	2.6107	2.6067	2.5995	2.5941	2.5873	166	201	131	1–15.5
TI1173	—	—	—	—	—	—	2.6083	2.6002	2.5962[c]	—	—	99	1.05–15
AMTIR3	—	—	—	2.6266	2.621	2.6173	2.6088	2.6083	2.5942	—	174	110	1.05–15
IG5	—	—	—	2.6277	2.6226	2.6187	2.6105	2.6038	2.5948	—	180	102	0.9–15.5
IG4	—	—	—	2.6263	2.621	2.6183	2.6121	2.6084	2.6029	—	203	176	0.8–12
IRG28	2.7394	2.7285	2.7276	2.712	2.7073	2.7026	2.694	2.6875	2.6788	146	162	111	1.2–13.5

IRG25	2.816	2.8081	2.8022	2.7894	2.7849	2.7804	2.7728	2.7675	2.7612	131	198	152	1.5–17
IG6	—	—	—	2.8014	2.7945	2.7907	2.7831	2.7775	2.7721	—	168	161	0.9–15.5
IG3n	—	—	—	2.8111	2.8034	2.7993	2.7919	2.787	2.781	—	153	164	1.5–15.5
BSA	—	—	—	—	—	—	2.7789	2.7758	2.7728	—	—	159	1–15
BS2	—	—	—	—	2.8732	2.8688	2.861	2.8563	2.8509	—	—	185	2–14
IRG32	3.0447	3.0351	3.0257	3.0072	2.9999	2.9926	2.981	2.9731	2.9635	107	137	113	1.5–15.5
AV1	—	3.0531	3.0463	3.0287	3.0204	3.0121	2.9996	2.992	2.983	—	122	120	—
AV2	—	2.8851	2.8782	2.8652	2.8582	2.8513	2.848	2.8424	2.8367	—	134	162	—
AV3	—	2.8021	2.7976	2.7692	2.7679	2.7667	2.7603	2.7514	2.7421	—	77	96	1–15.5
AV4	—	3.0758	3.0675	3.0433	3.0344	3.0256	3.0172	3.0128	3.01	—	115	279	1.2–17
IRG35	—	2.392	—	2.382	2.378	2.375	2.372	2.37	2.367	—	197	274	1.5–16
IRG27	—	—	—	—	—	2.688	2.679	2.673	—	—	—	—	2 to 4.2–16
IRG33	—	—	—	—	—	—	—	—	2.665	—	181	—	7.5–16.5
As$_2$S$_3$	—	2.4268[d]	—	2.4152	2.4116	2.4074	2.3937	2.3822	—	—	—	—	0.64–9

[a] $\lambda = 1.5$ μm
[b] $\lambda = 2.4$ μm
[c] $\lambda = 11$ μm
[d] $\lambda = 1.97$ μm

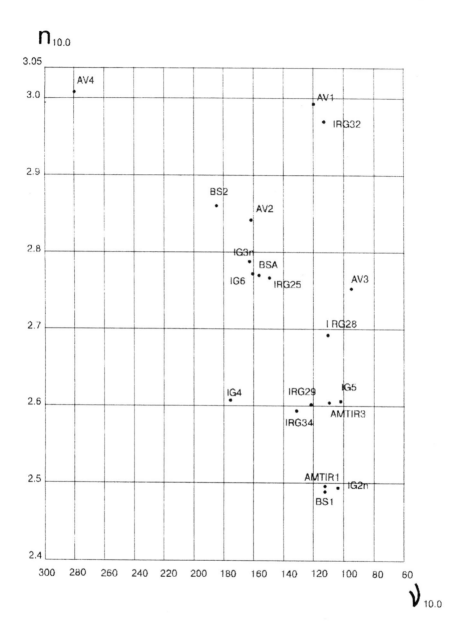

FIGURE 67.   Abbe diagram for commercial glasses of different firms in coordinates n_{10}–V_{10}.

20 to 120°C for the spectrum range 2 to 12 μm, in Tables 34 and 35. As could be expected, the glasses As_2S_3 and IRG23, AMTIR-3, and IRG34 have the closest value β.

Spectral curves of glass transmission are presented in Figures 68 to 71. The thickness of control samples is 10 mm, except the filters IRG27 and IRG33, the transmission coefficient was most frequently for a thickness of 3 mm.

TABLE 34
Absolute Temperature Coefficient of the
Refractive Index β_{abs} f the Glasses IRG

Glass type	$\beta_{abs}(t,\lambda) \times 10^{6}°C^{-1}$ for λ μm (20–120°C)						
	2	2.6	3.4	4.6	5	6–7	8–12
IRG23	9	8	7	7	7	7	—
IRG24	50	48	46	42	42	42	—
IRG25	62	55	50	46	46	46	46
IRG28	54	47	43	40	40	40	40
IRG29	58	52	49	44	44	44	44
IRG32	142	135	132	128	128	128	128
IRG34	109	103	97	96	96	96	96
IRG27	—	—	70	68	68	68	68

TABLE 35
Absolute Temperature Coefficients of the Refractive Index for the Glasses
AMTIR and As_2S_3

Glass type	Temperature interval (°C)	$\beta_{abs}(t,\lambda) \times 10^{6}°C^{-1}$ for λ μm						
		1.15	3.39	10.6	3	5	8	10	12
AMTIR1	25 → 65	101	77	72					
AMTIR3	25 → −197				58 ± 2	57 ± 2	55 ± 2	56 ± 2	56 ± 2
	25 → 150				98 ± 4	92 ± 4	87 ± 8	91 ± 11	93 ± 6
As_2S_3	25 → −78					−8.6			
	25 → 65					9.3			

For IRG25 the spectral curve is given of its own radiation at T = 160°C (Figure 69). Temperature changes of ChG transmissions are determined by the location of fundamental absorption bands. There is no absorption by free carriers which takes place in crystalline materials, including germanium; hence, in the transparency region the transmission doesn't change with the temperature.

The glass filters IRG27 and IRG33 (Figure 71) don't have analogues among commercial ChG. IRG27 filter with the shifting transmission edge in the region 1.7 to 3.7 μm is transparent to 16 μm. The location of the short-wave limit is ordered by a customer: IRG33 filter for the range 7.5 to 16.5 μm. The glass IRG28 can be used as a sublayer for a filter in the range 8 to 13.5 μm. Its long-wave transmission limit agrees with atmosphere transparency; the short-wave limit shift to 8 μm is achieved by reflecting and antireflection coatings. Integral transmission of this filter 8 to 10 mm thick in the range 8.0 to 12.0 μm is equal to 60%.

ChG clarification can be performed by two methods. The chemical clarification for the spectral range 0.6 to 10 μm increases the transmission of a component 23 to 30%.

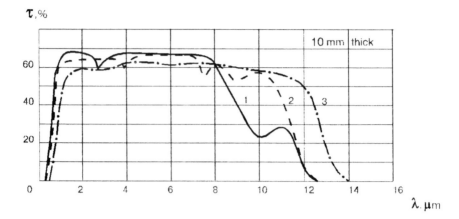

FIGURE 68.   Transmission spectra of IRG series glasses: (1) IRG23; (2) IRG24; (3) IRG28.

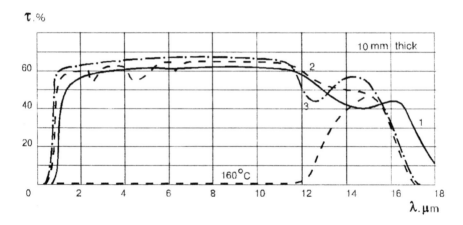

FIGURE 69.   Transmission spectra of IRG series glasses: (1) IRG25; (2) IRG29; (3) IRG34.

Resistive or electron-beam evaporation of clarifying substances in vacuum enhances the transmission coefficient in the spectrum range 8.0 to 14.0 μm to 0.9 to 0.98%.

Due to poor comparability and sometimes discrepancy of the results of the measurements of physicochemical properties of the glasses, produced by different companies, they are presented separately in Table 36 to 39. The glasses of the State Optical Institute are presented in three tables. In Table 37 mechanical and electrical properties of glasses of the series IRG are given; in Table 38 density, elasticity modulus, and thermal properties of the glasses IRG and AV. The viscosity characteristics of the glasses of the State Optical Institute are presented in Table 39. The technological process is elaborated according to the viscosity curves. The data about Tg (10^{13} P) are also received from them.

The property, common for all the glasses under study, is their high resistance against the attack of chemical reagents. They don't dissolve in water solutions of

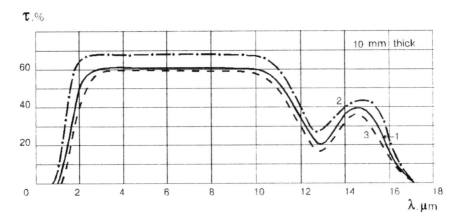

FIGURE 70.   Transmission spectra of IRG and AV glasses: (1) IRG32; (2) AV3; (3) AV4.

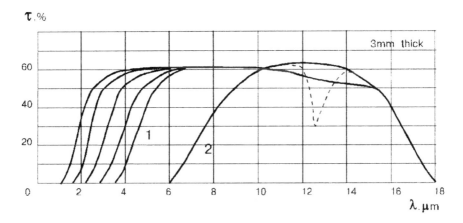

FIGURE 71.   Transmission spectra of filters: (1) IRG27; (2) IRG33.

acids and salts; hence they are resistant against the atmosphere moisture and sea-water, can't be destructed by hydrofluoric acid (unlike all oxide glasses), and don't change after being held in such organic solvents as gasoline, toluene, alcohols, or acetone.

ChG surface can be destroyed and affected by alkaline mediums, including ammonia. In optical glass-making, however, acid resistance isn't standardized. ChG retain their properties under the influence of penetrating radiation: after the irradiation of glasses with γ-ray 10^7 the transmission coefficient doesn't change in the working spectrum range.

The transmission quality of glasses is determined by the level of optical losses, expressed by the magnitude of the attenuation factor. Its values can be obtained in two ways: by direct measurement, for example, with the help of the laser calorimetric method, or by calculation from measurements of transmission coefficient and refractive

TABLE 36
Physicochemical Properties of the Glasses AMTIR, BS, IG

Glass type	ρ (g/cm³)	Tg °C	Tg K	Ta °C	Ta K	α × 10⁶ °C⁻¹	α × 10⁶ K⁻¹	E 10⁻⁶psi	E G-Pa	λ $\frac{cal}{S \times cm \times K} \times 10^4$	C cal/gK	H(Knoop) $\frac{Kg}{mm^2}$	H(Knoop) GPa	H(Vickers) $\frac{Kg}{mm^2}$	μ	p (Ω × cm)	T_use °C	T_use K
AMTIR1	4.4	362	—	370	—	12	—	3.2	—	6	0.07	170	—	—	0.27	2*10¹²	300	—
AMTIR3	4.67	278	—	285	—	13.5	—	3.11	—	5.3	0.066	150	—	—	0.26	5*10¹¹	200	—
AS₂S₃	3.2	180	—	208	—	21.4	—	2.3	—	4	0.109	109	—	—	0.24	—	150	—
BS1	4.42	351[146]	612	—	630	—	12.8	—	—	—	—	180	—	245	—	—	—	540
BS2	4.9	262[146]	517	—	537	—	12.8	—	—	—	—	—	—	226	—	—	—	470
IG2n	4.41	368	—	—	—	—	13	—	21.5	—	—	—	1.4	—	0.25	—	—	—
IG3n	4.48	279	—	—	—	—	14.1	—	22	—	—	—	1.36	—	0.21	—	—	—
IG4	4.47	225	—	—	—	—	19.4	—	—	—	—	—	1.12	—	—	—	—	—
IG5	4.66	285	—	—	—	—	12.8	—	—	—	—	—	1.13	—	—	—	—	—
IG6	4.63	185	—	—	—	—	19.6	—	—	—	—	—	—	—	—	—	—	—

Note: ρ — density; T_g — glass transition temperature; T_a — annealing temperature; λ — coefficient of the thermal expansion; E — Young's modulus, λ — thermal conductivity; C — specific heat; H — microhardness; μ — Poisson's ratio, p — resistivity; T_{use} — upper use temperature.

TABLE 37
Mechanical and Dielectrical Properties of the Glasses IRG

Glass type	$\sigma \times 10^{-5}$ P	μ	$B \times 10^{12}$ Pa⁻¹ ($\lambda = 2$ μm)	$H \times 10^{-7}$ Pa	H_s	$\varepsilon(f,t)$ at 9547 MHz	$tg\delta \times 10^4$ at 9547 MHz	ρ Ω × cm 20°C	ρ Ω × cm 50°C	ν_λ ang min × A⁻¹ ($\lambda = 1.1523$ μm)
IRG23	196	0.31	−10.2	147	0.13	7.9	2.8	1.2×10^{17}	—	0.065
IRG24	196	0.28	−3.26	196	0.11	7.5	1.6	2.5×10^{14}	—	0.151
IRG25	196	0.29	−31.62	147	0.12	10.3	6.5	3.2×10^{11}	1.3×10^{7}	0.17
IRG28	196	0.29	−27.03	147	0.14	9.9	5.5	8.2×10^{12}	6×10^{7}	0.12
IRG29	196	0.28	−22.44	157	0.15	9.8	5	4×10^{13}	2.5×10^{8}	0.122
IRG32	196	0.26	−20.4	206	0.12	22.1	6.7	8.9×10^{9}	—	0.12
IRG34	245	0.24	−10	245	0.15	9.4	6.5	—	—	0.116
IRG27	196	0.28	—	186	0.14	11.6	29	3.3×10^{15}	2×10^{9}	—
IRG33	196	0.28	—	147	0.14	10.8	36	—	—	—

Note: σ—bending strength; μ—Poisson ratio; B—Brewster constant; H—microhardness; H_s—relative microhardness to grinding; ε—dielectric penetrability; $tg\delta$—angle's tangents of dielectric losses; ρ—specific ohmic resistance; ν_λ—magneto-optical constant (Verdet constant).

TABLE 38
Density, Elasticity Modulus, and Thermal Properties of the Glasses IRG

Glass Type	ρ (g/cm³)	T_g (°C)	T_{ds} (°C)	$\alpha_t \times 10^6$°C⁻¹ From −60 to +20°C	$\alpha_t \times 10^6$°C⁻¹ From +20 to T_g	$E \times 10^{-7}$ Pa	λ (W × m⁻¹ × °C⁻¹)	c (J × kg⁻¹°C⁻¹)	$a \times 10^8$ m² × s	T_{use} (°C)
IRG23	3.32	162	200	20.5	24.6	1569	0.333	0.46	19	100
IRG24	3.89	239	278	16.2	18.2	1868	0.375	0.54	20	160
IRG25	4.72	156	190	19.7	22	1868	0.343	0.38	21	100
IRG28	4.43	167	200	19.7	22	1764	0.347	0.36	21	110
IRG29	4.74	173	200	18.3	22	1828	0.368	0.41	20	110
IRG32	5.66	238	250	13.3	14.7	2715	0.275	0.3	16.5	190
IRG34	4.47	338	375	11.4	12.5	2254	0.3	—	—	250
IRG27	4.89	237	270	—	17.7	1960	0.275	0.33	17	170
IRG33	4.97	145	195	—	22.5	2009	0.28	—	—	80
AV1	4.94	209	238	—	14.3	—	—	—	—	120
AV2	4.84	286	311	—	13.1	—	—	—	—	230
AV3	4.88	234	267	—	13.8	—	—	—	—	175
AV4	5.05	130	165	—	23.5	—	—	—	—	75

Note: ρ—density; T_g—glass transition temperature 10^{13} P; T_{ds}—dilatometrical softening temperature; α—thermal coefficient of expansion; E—elasticity modulus; λ—thermal conductivity; c—specific heat; a—temperature conductivity; T_{use}—upper use temperature.

TABLE 39
Temperature Dependence of the Glasses IRG and AV Viscosity

| Glass | Temperature at lgη (P)°C | | | | | | | | | | |
type	13	12	11	10	9	8	7	6	5	4	3
IRG23	162	175	190	205	220	237	256	279	309	347	374
IRG24	239	256	274	295	319	346	378	415	460	507	580
IRG25	156	170	183	198	214	232	253	275	305	336	377
IRG28	167	179	193	206	221	239	258	281	309	339	378
IRG29	173	190	207	223	239	257	276	302	332	368	396
IRG32	238	248	257	268	281	294	306	330	—	—	—
IRG34	338	360	382	405	428	451	473	498	527	553	580
IRG27	237	252	268	286	304	324	348	375	407	440	482
IRG33	145	166	187	209	230	254	—	—	—	—	—
AV1	209	222	234	247	259	277	294	310	362	—	—
AV3	234	249	264	280	295	310	334	356	410	—	—
AV4	130	146	161	175	191	210	233	261	304	—	—

index. The spectral transmission coefficient τ_λ is determined as the relation of the radiation flow ϕ_λ, passed through the glass, to the falling flow ϕ_0, i.e., $\tau_\lambda = \phi_\lambda/\phi_0$. The value $1 - \tau_\lambda$ characterizes the complete radiation losses, determined by attenuation and reflection from the polished surfaces of a detail. Attenuation factor ε_λ cm^{-1} is determined by the formula $\varepsilon_\lambda = (D_\lambda - D_{\mu m})/1$, where $D_\lambda = -lg\tau_\lambda$ is optical density; $D_{\mu m} = -lg[2n_\lambda/n_\lambda^2 + 1]$ is the correction for the repeated reflection from two surfaces; 1 is the thickness of a sample in the direction of beam's way, cm. The absorption coefficients of the glasses, manufactured by the company Amorphous Materials, obtained by direct measurements, vary in the region from 2.0 to 10 μm in the limits 0.001 to 0.008 cm^{-1} and this answers the requirements of the glasses of laser and fiber optics (Table 40). For the application in the apparatuses for thermal vision etc., the sufficient absorption level is 0.01 to 0.02 cm^{-1}, which is provided in all commercial IR glasses.

Attenuation factors of the glasses of the State Optical Institute are calculated from the formula noted above. In Table 41 the values of the transmission coefficient, attenuation index (ε_λ/cm^{-1}), and corrections for the repeated reflection of standardized glasses are given. One can see that in the ranges of maximum transparency ε_λ are equal to not more than 0.01 to 0.03 cm^{-1} for different types of glasses. According to a customer's order the glass IRG34 can be prepared with losses 0.001 cm^{-1}.

For the glasses filters the deviations of short-wave transmission edge are standardized as follows: for IRG27 $\lambda_e = (\lambda_{giv} \pm 0.1)$ μm, where λ_{giv} is any value in the limits 1.7 to 3.7 μm, prescribed by a consumer; λ_e for IRG33 = (7.5 ± 0.5) μm. The transmission edge is characterized by the wavelength at which the transmission coefficient is one-half of its highest value in the working part of the spectrum.

Optical constants and optical homogeneity are standardized by different companies in different ways. The company Amorphous Materials guarantees the changes

TABLE 40
Absorption Coefficient K cm^{-1}/λμm

	0.6439	1.014	1.53	1.907	3	4	5	6	7	8	9	10	11	12	13	14
As$_2$S$_3$	0.42	0.02	0.015	0.01	0.02	0.014	0.003	0.003	0.007	0.02	0.31	0.85	0.85	—	—	—
AMTIR1	—	0.07	0.01	0.004	0.003	0.002	~0.001	~0.001	~0.001	~0.001	0.006	0.008	0.03	0.15	0.15	0.13
AMTIR3	—	—	—	—	0.002	~0.001	~0.001	~0.001	~0.001	0.002	0.004	0.008	0.03	0.13	0.2	0.2
BS1	—	—	0.15	0.01	—	0.01	—	0.01	—	0.01	—	0.03	0.03	0.15	—	—
BS2	—	—	—	—	—	—	—	—	—	—	—	10.6	11.6	12.6	—	—
												0.03	0.08	0.03		

TABLE 41
Values of the Transmission Coefficient τ_λ, Attenuative Index $\epsilon_\lambda (cm^{-1})$, Corrections for the Repeated Reflection from Both Surfaces of the Article $D_{\rho m}$

Glass type	λ μm	$\tau\lambda$	$\epsilon\lambda$ (cm^{-1})	$D_{\rho m}$
IRG23	1.4	0.66	0.02	0.156
	2	0.68	0.02	0.154
	3	0.6	0.07	0.151
	4–7.5	0.68	0.02	0.15
	8	0.65	0.06	0.148
	9	0.42	0.23	0.148
IRG24	1.4	0.63	0.05	0.152
	2–3.5	0.65	0.05	0.15
	4	0.6	0.07	0.148
	4.5–7	0.67	0.03	0.148
	7.8	0.55	0.11	0.146
	10	0.58	0.09	0.145
IRG25	2	0.56	0.05	0.199
	3	0.6	0.02	0.197
	4–12	0.61	0.01	0.195
	14	0.43	0.16	0.193
	16	0.43	—	—
IRG28	2	0.6	0.03	0.19
	4–8	0.63	0.02	0.186
	9	0.6	0.04	0.185
	10	0.56	0.07	0.185
	11	0.53	0.09	0.184
	12	0.5	0.12	0.184
IRG29	2	0.6	0.04	0.178
	2.8	0.57	0.07	0.177
	3.5	0.62	0.03	0.176
	4.5	0.57	0.07	0.176
	5.5	0.65	0.01	0.175
	6.3	0.62	0.03	0.175
	7–11	0.65	0.01	0.174
	12	0.6	0.05	0.173
	13–14	0.51	0.13	0.172
IRG32	2	0.53	0.05	0.226
	3–10	0.6	0.01	0.22
	11	0.56	0.03	0.216
	12	0.35	0.24	0.216
	14	0.4	—	—
IRG34	1.4	0.6	0.04	0.18
	2	0.62	0.03	0.177
	3–11	0.65	0.01	0.174
	12.5	0.45	0.17	0.172
	14	0.55	0.09	0.171

TABLE 41 (continued)
Values of the Transmission Coefficient τ_λ, Attenuative
Index $\epsilon_\lambda(cm^{-1})$, Corrections for the Repeated
Reflection from Both Surfaces of the Article $D_{\rho m}$

Glass type	λ μm	$\tau\lambda$	$\epsilon\lambda$ (cm^{-1})	$D_{\rho m}$
IRG27	6–10	0.63	0.06	0.184
IRG33	10–12	0.6	0.15	0.183
	13	0.5	0.02	0.182

of refractive index from batch to batch less than: AMTIR-1 = ±0.001; AMTIR-3 = ±0.005; As$_2$S$_3$ = ±0.003. Optical homogeneity of AMTIR-1 for a 16.2 mm-diameter plate 26 mm thick was ±20 × 10^{-6} or $\Delta n/n = 8 \times 10^{-6}$. AMTIR-3 for a 159 mm diameter plate 21 mm thick was ±50 × 10^{-6} or $\Delta n/n = 19 \times 10^{-6}$. Barr and Stround Company doesn't present standardized values of optical constants; as for homogeneity and absence of insertions, they are guaranteed by technology.

The company "LZOS", which produces glasses for the State Optical Institute, considers it necessary to standardize the series IRG according to all quality parameters accepted in optical glass-making. In accordance with it, along with weakening rate, the refractive index and dispersion are standardized. For the glasses IRG23, IRG24, IRG25, IRG28, IRG29, IRG32, and IRG34 the threshold deviation of the refractive index $\Delta n_{2.0}$, $\Delta n_{4.0}$, and $\Delta n_{10.0}$ from the values, given in Table 33, is ±30 × 10^{-4}. The threshold deviation of dispersion $\Delta(n_{1.8}-n_{2.2})$, $\Delta(n_{3.0}-n_{5.0})$, $\Delta(n_{8.0}-n_{12.0})$ is equal to ±50 × 10^{-5}. On the customer's demand batches of glasses can be prepared with deviations $\Delta n = ±10 \times 10^{-4}$, $\Delta v = ±30 \times 10^{-5}$.

Double refraction — The permissible values of double refraction in the working direction of a preparation are presented in Table 42. According to this parameter the glasses are divided into three categories.

Striaes — In the preparation of glasses for the details working at $\lambda = 5$ μm striaes are not allowed, which introduces into the wave front the path difference more than 0.1 of a working wavelength (I category) and 0.2 (II category).

Bubbles and insertions — The permissible number of bubbles and insertions with diameter 10 to 100 μm does not exceed 1000 on 1 cm^3 of glass. The insertions with diameter more 100 μm are not allowed. Selective control of double refraction, is possible subject to customer's order nonstriaeness, and presence of bubbles. The standardized quality of glass in conformity with these parameters must be provided by the technological process. In OST 3–3441–83 devices and method for the control over all quality parameters are noted. T_{use} for the glasses produced in the State Optical Institute was determined from viscosity characteristics (Table 39) as the temperature, for the viscosity 10^{18} P. As the experimental viscosity values can be obtained by the impression method only for the interval 10^{13} to 10^4 P, the temperature for the greater viscosity values is calculated by the Fulcher formula:

TABLE 42
Values of Birefringence of Glasses in Working Direction

Glass type	Birefringence at λ_2 μm, nm/cm		
	Category 1	Category 2 no more than	Category 3
IRG23	60	150	300
IRG24	20	50	100
IRG25	200	500	1000
IRG28	150	400	800
IRG29	150	350	700
IRG32	130	300	600
IRG34	60	150	300

$$\lg \eta = A + \frac{B}{T - T_0}$$

The coefficients A, B, T_0 were calculated through the calculation of the system of three equations for their known values $\lg \eta$, equal to 13, 9, and 5. The temperature, for the viscosity 10^{18} P, was determined from the plot $\lg \eta - 1/T$, extrapolated toward the high viscosity values. Its rounded magnitude is standardized as the upper use temperature for this glass exploitation. The dimensions of the glass preparations, manufactured by different companies, are given in Table 43.

Figures 72 and 73 show examples of articles manufactured by the Joint-Stock Company "LZOS".

Unfortunately, we haven't gotten any information about the quality of glasses of the IG series.

4.5   PROBLEMS IN THE APPLICATION OF OPTICAL CHALCOGENIDE GLASSES

As was said in Section 4.1, in the last years, owing to the attention which has been paid to the ecological purity of materials, the application of ChG in IR optics has been reduced because of false ideas about their toxicity connected with presence of arsenic in the compositions. Arsenic oxidation to poisonous dioxide takes place, however, at a temperature not lower than 300°C; hence it can't be formed in the conditions of application.

There exist certain safety measures during manufacturing, for ChG synthesis is carried out at high temperatures on all stages; in addition, workers are constantly in contact with dust and powders of initial substances. The accepted safety measures

TABLE 43
Largest Sizes of Products of the Different Companies

Glass type	Form and size of the products
AMTIR1, AMTIR3	Plates up to 30 cm × 45 cm
As_2S_3	Blanks up to 8 in. diameter or plates 30 cm × 45 cm (12 × 18 in)
BS1, BS2	Blanks up to 160 mm in diameter up to 30 mm thick,
	75 mm in diameter 4 to 5 mm thick, 30 mm in diameter to 1 mm thick
IRG23, IRG24, IRG25,	Disks up to 350 mm in diameter
IRG28, IRG29, IRG34	
IRG32	Disks up to 150 mm in diameter
AV1, AV2, AV3, AV4	Disks up to 250 mm in diameter
	Diameter/thickness
	ratio is from 3:1 to 10:1
IRG27	Disks up to 100 mm in diameter
IRG33	Disks up to 50 mm in diameter

Note: All glasses can be manufactured as optical elements of various configurations, any shape or size.

FIGURE 72. Different forms of IRG glass articles.

are, apparently, sufficient, because after more than 30 years of industrial manufacturing of these glasses, no symptoms of poisoning among glassmakers have been noted. Even in the 1960s and 1970s, when these materials were widely applied in our country and dozens of tons were produced annually, no claims were heard either from manufacturers or from consumers.

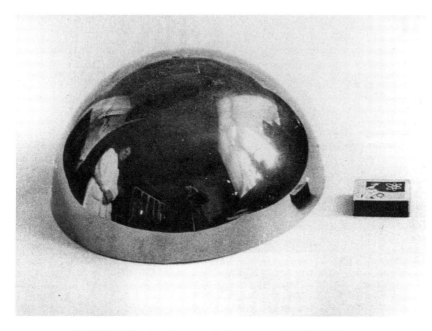

FIGURE 73.   Semisphere of glass made of IRG25 250 mm.

The standard technological process for ChG cold treatment, including the safety measures instruction, was elaborated in the State Optical Institute.[220] It is considered that comparatively low mechanical strength and thermal stability prevent their application, but therewith their much higher "elasticity", in comparison with other IR materials, isn't, as a rule, taken into account. Young's modulus in ChG is considerably lower, which aligns CTE high values, and if one takes into account that the details inside the devices are held in isothermal conditions, the temperature interval from –60 to +60°C accepted when testing articles appears to be quite sufficient. "LZOS" as a rule supplies the material in half-finished articles (Figures 72, 73). Cold treatment and coating application are performed by a customer. Drastic thermal shocks and heating of articles above T_{use} should be avoided. Elementary safety measures at ChG synthesis and treatment are reported in Chapter 1.

Some companies pay special attention to the small surface hardness of ChG. This is really so, but in modern devices antireflecting, deflecting, and other types of coatings are applied on the surfaces of articles, hence their abrasion resistance is determined by the coating's lasting. The mechanical properties of ChG, naturally, predetermine the necessity of delicate handling of articles when fitted into metal rims: it's necessary to use therewith any highly elastic material, vacuum-rigid enough and with good adhesion to metal and glass.

Experiment shows that properly exploited ChG can be used in IR optics for several years, without any changes in the article's quality. Several examples confirm this. Thus, the glass IRG23, as was already mentioned, has replaced crystalline material as the thermal vision objective of a mass-produced device, working in atmospheric windows I and II. The number of lens in the system have therewith decreased from eight to three, with no reduction in the device parameters. IRG24 is applied in illuminators and optical articles for photo devices, installed on low-speed airplanes — in particular, in the form of plates $150 \times 150 \times 8$ mm, with six plates in each set. Two deflectors for self-guidance caps of the apparatus, which works in IR and radio ranges, were also made of IRG24. The deflectors were cone-shaped Ø500 mm, 1000 mm high, and they were agglutinated from four collars of pyramid-shaped plates — 16 plates in each collar. The articles were put through natural testings and demonstrated good transparency in both ranges. Glasses IRG23 and IRG24 were also used as lens Ø60 mm in a device for the investigation of sun radiation with parachute landing; in these conditions therewith their rigidity appeared to be quite sufficient. Filters made of IRG28 Ø108 mm by 8 mm work successfully in space as outside windows-lens of the device, installed on the satellites of the "Meteor" series. With the help of this device high-quality photos were taken of the European continent, which were demonstrated on the exhibition of the achievements of national economy. As a customer reported, the glass safely withstands the device's transfer from the sunny to the shadow side and back. Glass IRG25 is applied as the windows for bolometers and in the objectives of precision devices, which use the third atmospheric window, in particular, as thin lens, coupled with lens made of optical ceramics.

Glasses IRG29 and IRG34 are applied as windows in lasers on CO_2 low and middle power. Glass IRG29 is also used as lens and prisms — in particular, telescopic systems and, in particular, as the Dove prism. The protective windows for instrumentation of various functions were made of this glass, in particular, the protective window Ø370 mm in a particular article, working in the conditions of raised dampness and vibration. In the same conditions and also as an entrance window the glass IRG25 worked in a device, which measured with a high accuracy the temperature difference in seawater. The window more than 200 mm across, installed at an angle, rotated with the device at a speed of 160 rpm and was attacked by the splashing seawater.

These examples of the successful application of the least heat-resistant and thermo-rigid glasses in difficult working conditions demonstrate that in cases when the requirements of the transparency range predominate, or to the combination of optical constants, thermal and mechanical properties shouldn't stop the constructors.

Chalcogenide optical glasses are still very young — less than 40 years have passed since the day of their creation. This is a very short period for any commercial material, and the author of this monograph is absolutely sure that the future of infrared optics is in chalcogenide glasses.

TABLE 1

		Properties of glasses As–Ge–Se, Sb–Ge–Se, Sb–Sn–Ge–Se, As–Sn–Ge–Se, As–Bi–Ge–Se					
N_0N_0	Composition of glass (at.%)	T_{ds} (°C)	$\alpha(10^6 \ °C^{-1})$	H (kg/mm²)	T_g (°C) 10^{13} P	ΣK_{gl}	$\Sigma\varepsilon_{gl}$
1	2	3	4	5	6	7	8
1	As_5Se_{95}	—	—	67	47	1.025	50.7
2	$As_{10}Se_{90}$	77	42	80	58	1.05	52.3
3	Ge_5Se_{95}	—	—	80	57	1.05	52.4
4	$As_{15}Se_{85}$	85	37.5	90	71	1.075	54
5	$As_5Ge_5Se_{90}$	90	36.8	90	78	1.075	54.3
6	$As_{20}Se_{80}$	97	33.5	105	—	1.1	55.7
7	$Ge_{10}Se_{90}$	—	—	105	83	1.1	55.8
8	$As_{10}Ge_5Se_{85}$	101	33.8	105	88	1.1	55.8
9	$As_{25}Se_{75}$	115	31.5	116	96	1.125	57.4
10	$As_{15}Ge_5Se_{80}$	119	32.5	119	—	1.125	57.5
11	$As_5Ge_{10}Se_{85}$	—	—	116	—	1.125	57.6
12	$As_{15}Sn_5Se_{80}$	—	—	120	104	1.125	57.2
13	$Sb_{10}Ge_{7.5}Se_{82.5}$	120	32.5	116	—	1.125	57.2
14	$As_{30}Se_{70}$	136	28	130	—	1.15	59.1
15	$As_{20}Ge_5Se_{75}$	132	29	131	—	1.15	59.1
16	$As_{10}Ge_{10}Se_{80}$	132	28.5	130	—	1.15	59.2
17	$Sb_{15}Ge_{7.5}Se_{77.5}$	136	28.7	119	—	1.15	58.8
18	$Sb_{10}Sn_5Ge_5Se_{80}$	130	31.5	—	—	1.15	58.3
19	$Ge_{15}Se_{85}$	140	31.3	130	—	1.15	59.4
20	$Sb_{10}Sn_{7.5}Ge_5Se_{77.5}$	133	30	—	—	1.175	59.8
21	$As_{25}Ge_5Se_{70}$	160	26.2	145	125	1.175	60.8
22	$As_{15}Ge_{10}Se_{75}$	156	27.5	146	123	1.175	60.9
23	$As_5Ge_{15}Se_{80}$	159	26.8	144	133	1.175	61
24	$Sb_5Ge_{15}Se_{80}$	158	27.2	138	—	1.175	60.8
25	$Sb_{15}Ge_{10}Se_{75}$	155	27.5	133	—	1.175	60.5
26	$Sn_{7.5}Ge_{12.5}Se_{80}$	163	28.5	144	124	1.175	60.8
27	$Sb_{10}Sn_{7.5}Ge_{7.5}Se_{75}$	166	25.2	—	—	1.2	61.5
28	$Sn_5Ge_{15}Se_{80}$	169	28.2	147	133	1.2	61.6
29	$Sb_{10}Sn_5Ge_{10}Se_{75}$	165	25.3	—	—	1.2	61.8
30	$Ge_{20}Se_{80}$	—	—	160	157	1.2	62.8
31	$As_{30}Ge_5Se_{65}$	177	23.5	160	152	1.2	62.5
32	$As_{20}Ge_{10}Se_{70}$	170	24.5	162	157	1.2	62.6
33	$Sb_{10}Ge_{15}Se_{75}$	175	25	146	151	1.2	62.4
34	$Sb_{15}Ge_{12.5}Se_{72.5}$	175	24.9	139	—	1.2	62.3
35	As_2Se_3	195	21	160	170	1.2	62.4
36	$As_{45}Se_{55}$	192	23.6	150	160	1.225	62.8
37	$Sb_5Sn_5Ge_{15}Se_{75}$	189	25	—	—	1.225	63.7
38	$Sb_5Sn_{10}Ge_{10}Se_{75}$	185	25.5	—	—	1.225	63.4
39	$Sb_{10}Sn_5Ge_{12.5}Se_{72.5}$	183	24	—	—	1.225	63.5
40	$As_{50}Se_{50}$	182	26	122	167	1.25	63.5

TABLE 1 (continued)

N_0N_0	Composition of glass (at.%)	T_{ds} (°C)	$\alpha(10^6 \ °C^{-1})$	H (kg/mm²)	T_g (°C) 10^{13} P	ΣK_{gl}	$\Sigma \varepsilon_{gl}$
1	2	3	4	5	6	7	8
	Properties of glasses As–Ge–Se, Sb–Ge–Se, Sb–Sn–Ge–Se, As–Sn–Ge–Se, As–Bi–Ge–Se						
41	$As_{35}Ge_5Se_{60}$	215	20.8	168	182	1.225	64
42	$As_{25}Ge_{10}Se_{65}$	202	21.5	169	176	1.225	64.2
43	$As_{20}Ge_{12.5}Se_{67.5}$	200	22.6	—	—	1.225	—
44	$As_5Ge_{20}Se_{75}$	205	22.6	176	173	1.225	64.4
45	$Sb_{15}Ge_{15}Se_{70}$	200	22.4	153	—	1.225	64
46	$Sb_{10}Sn_{10}Ge_{10}Se_{70}$	206	21	—	—	1.25	64.6
47	$As_{45}Ge_5Se_{50}$	209	24.4	146	180	1.275	65
48	$Sb_{10}Sn_5Ge_{15}Se_{70}$	210	21	—	—	1.275	65.2
49	$Sb_{10}Ge_{20}Se_{70}$	219	21	166	200	1.275	65.5
50	$Sb_{15}Ge_{17.5}Se_{67.5}$	222	20	161	—	1.275	65.7
51	$Sn_{7.5}Ge_{17.5}Se_{75}$	226	21.5	170	186	1.25	65.2
52	$As_{30}Ge_{10}Se_{60}$	238	19.3	174	212	1.25	66.5
53	$As_{20}Ge_{15}Se_{65}$	236	20	183	212	1.25	66
54	$As_{10}Ge_{20}Se_{70}$	230	20.8	192	202	1.25	66.1
55	$Ge_{25}Se_{75}$	—	—	184	214	1.25	66.2
56	$Sb_{10}Sn_{7.5}Ge_{15}Se_{67.5}$	243	18.7	—	—	1.275	66.7
57	$Sb_{10}Sn_{10}Ge_{12.5}Se_{67.5}$	234	18.8	—	—	1.275	66.4
58	$Sb_{15}Sn_5Ge_{15}Se_{65}$	248	18	—	—	1.275	66.8
59	$As_{40}Ge_{10}Se_{50}$	231	21	167	211	1.3	66.6
60	$As_{55}Ge_5Se_{40}$	236	23	154	205	1.325	66
61	$As_{60}Ge_5Se_{35}$	—	—	174	211	1.35	66.5
62	$As_{35}Ge_{10}Se_{55}$	—	—	170	211	1.275	66
63	$As_{25}Ge_{15}Se_{60}$	265	18.2	186	233	1.275	67.1
64	$As_{15}Sn_5Ge_{15}Se_{65}$	—	—	191	237	1.275	67.2
65	$Sb_{15}Ge_{20}Se_{65}$	275	17.2	165	—	1.275	67.4
66	$As_{15}Bi_5Ge_{20}Se_{60}$	283	16.4	199	248	1.3	67.8
67	$As_{30}Ge_{15}Se_{55}$	275	17.6	188	241	1.3	67.6
68	$As_{45}Ge_{10}Se_{45}$	—	—	192	228	1.325	67.2
69	$As_{50}Ge_{10}Se_{40}$	—	—	202	237	1.35	67.5
70	$As_{15}Ge_{20}Se_{65}$	283	17.8	194	252	1.275	67.8
71	$As_{35}Ge_{15}Se_{50}$	277	18.3	192	243	1.325	68.2
72	$As_{25}Bi_5Ge_{20}Se_{50}$	302	14.4	210	258	1.35	68.9
73	$As_{40}Ge_{15}Se_{45}$	290	17	213	255	1.35	68.6
74	$As_{20}Ge_{20}Se_{60}$	294	17.3	192	266	1.3	68.6
75	$As_{25}Ge_{20}Se_{55}$	296	17.4	200	263	1.325	69
76	$As_{45}Ge_{15}Se_{40}$	300	15.5	256	262	1.375	69.2
77	$As_{30}Ge_{20}Se_{50}$	307	15.6	215	275	1.35	69.6
78	$As_{50}Ge_{15}Se_{35}$	317	13.8	245	287	1.4	69.6
79	$As_{15}Ge_{25}Se_{60}$	325	16	208	299	1.375	70.1
80	$As_{35}Ge_{20}Se_{45}$	330	14.4	232	285	1.375	70.1
81	$As_{15}Sn_5Ge_{25}Se_{55}$	—	—	227	284	1.375	70.5
82	$As_{20}Ge_{25}Se_{55}$	—	—	216	296	1.35	70.6

TABLE 1 (continued)

Properties of glasses As–Ge–Se, Sb–Ge–Se,
Sb–Sn–Ge–Se, As–Sn–Ge–Se, As–Bi–Ge–Se

N_0N_0 1	Composition of glass (at.%) 2	T_{ds} (°C) 3	$\alpha(10^6$ °C$^{-1})$ 4	H (kg/mm^2) 5	T_g (°C) 10^{13} P 6	ΣK_{gl} 7	$\Sigma \varepsilon_{gl}$ 8
83	As$_{45}$Ge$_{20}$Se$_{35}$	—	—	269	322	1.425	71.1
84	As$_{25}$Ge$_{25}$Se$_{50}$	342	14	237	317	1.375	71.2
85	As$_{30}$Ge$_{25}$Se$_{45}$	351	13.2	245	319	1.4	71.5
86	As$_{10}$Ge$_{30}$Se$_{60}$	364	14.2	219	333	1.35	71.8
87	As$_{30}$Ge$_{30}$Se$_{40}$	—	—	269	336	1.45	72
88	As$_{15}$Ge$_{30}$Se$_{55}$	370	13.6	233	338	1.375	72
89	As$_{20}$Ge$_{30}$Se$_{50}$	375	12.6	245	341	1.4	72.5
90	As$_{27.5}$Ge$_{27.5}$Se$_{45}$	369	12.4	259	—	1.412	72.2
91	As$_{35}$Ge$_{25}$Se$_{40}$	—	—	269	336	1.425	72
92	As$_{40}$Ge$_{25}$Se$_{35}$	—	—	282	342	1.45	72.4
93	As$_5$Ge$_{35}$Se$_{60}$	379	12.4	230	—	1.375	73.6
94	As$_{10}$Ge$_{35}$Se$_{55}$	379	12.4	245	—	1.4	73.5
95	As$_{25}$Ge$_{30}$Se$_{45}$	380	11.6	259	351	1.425	73
96	As$_{30}$Ge$_{30}$Se$_{40}$	387	10.7	269	361	1.45	73.4
97	As$_{2.5}$Ge$_{37.5}$Se$_{60}$	379	12.6	237	—	1.387	73.5
98	As$_{15}$Ge$_{35}$Se$_{50}$	380	11.5	260	358	1.425	73.9
99	As$_{20}$Ge$_{35}$Se$_{45}$	—	—	273	362	1.45	74.3
100	As$_{35}$Ge$_{30}$Se$_{35}$	—	—	286	372	1.475	74
101	As$_{40}$Ge$_{30}$Se$_{30}$	—	—	296	383	1.5	74.4
102	As$_{25}$Ge$_{35}$Se$_{40}$	—	—	286	377	1.475	74.9
103	As$_{30}$Ge$_{35}$Se$_{35}$	—	—	307	386	1.5	75.3
104	As$_{35}$Ge$_{35}$Se$_{30}$	413	—	—	398	1.525	75.7
105	As$_{40}$Ge$_{35}$Se$_{25}$	—	—	329	406	1.55	76.1

TABLE 2
Properties of Glasses As–Ge–Se–Sn–Pb.

Composition of glasses, at.%					T_{ds}	ρ	
As	Ge	Se	Sn	Pb	(°C)	(g/cm³)	N*10³
22	30	48	—	—	375	4.492	58.99
22	29.75	48	0.25	—	368	4.486	58.82
22	29.5	48	0.5	—	355	4.489	58.76
22	29.25	48	0.75	—	347	4.502	58.85
22	29	48	1	—	340	4.512	58.89
22	28	48	2	—	333	4.56	59.16
22	26.5	48	3.5	—	324	4.631	59.55
22	25	48	5	—	315	4.69	59.77
22	24	48	6	—	314	4.751	60.19
22	29.75	48	—	0.25	357	4.518	59.06
22	29.5	48	—	0.5	355	4.539	59.08
22	29.25	48	—	0.75	352	—	—
22	29	48	—	1	350	4.591	59.23
22	28	48	—	2	345	4.672	59.25
22	26.5	48	—	3.5	336	4.795	59.29
22	25	48	—	5	327	4.921	59.35
22	24	48	—	6	324	5.012	59.47
22	23	48	—	7	327	5.088	59.47
22	28.5	48	1	0.5	333	4.59	59.38
22	28	48	1	1	336	4.639	59.5
22	27.5	48	1	1.5	339	4.685	59.58
22	25	48	1	4	333	4.886	59.85
22	24	48	1	5	331	4.979	59.75
22	20	48	1	9	272	5.361	60.41
22	19	48	1	10	—	—	—
22	29.5	48	2	0.5	330	4.56	59.16
22	27	48	2	1	330	4.567	59.38
22	26.5	48	2	1.5	336	—	—
22	26	48	2	2	338	4.745	59.49
22	25	48	2	3	334	4.837	59.63
22	24	48	2	4	326	—	—
22	27	48	2	5	322	5.024	59.94
22	20	48	2	8	265	5.334	60.71
22	18	48	2	10	260	5.418	59.84
22	28	48	1	1	336	4.64	59.5
22	27.5	48	1.5	1	331	—	—
22	27	48	2	1	329	4.665	59.49
22	25	48	4	1	319	—	—
22	24	48	5	1	313	4.801	60.15
22	23	48	6	1	312	—	—
13	32	55	—	—	348	4.439	58.1
13	31	55	1	—	348	4.469	58.27
13	30	55	2	—	339	4.507	58.29

TABLE 2 (continued)
Properties of Glasses As–Ge–Se–Sn–Pb.

Composition of glasses, at.%					T_{ds}	ρ	
As	Ge	Se	Sn	Pb	(°C)	(g/cm³)	N*10³
13	27	55	5	—	318	4.591	58.34
13	26	55	6	—	312	4.6	58.11
13	31	55	—	1	346	4.489	52.87
13	30	55	—	2	342	4.59	58.03
13	27	55	—	5	335	4.821	57.99
13	26	55	—	6	333	4.88	57.77
13	19	55	—	13	273	—	—
13	17	55	—	15	260	—	—
13	30	55	1	1	337	4.5513	58.19
13	29.5	55	1	1.5	338	—	—
13	27	55	1	4	330	4.779	58.11
13	25	55	1	6	325	4.947	58.21
13	22	55	1	9	—	5.269	59.21
13	29	55	2	1	332	4.576	58.17
13	28.5	55	2	1.5	335	4.636	58.43
13	28	55	2	2	336	4.686	58.55
13	27	55	2	3	333	—	—
13	25	55	2	5	320	4.954	58.94
13	22	55	2	8	270	5.164	58.62
13	20	55	2	10	260	5.358	59.51
13	30	55	1	1	337	4.551	58.19
13	29.5	55	1.5	1	333	—	—
13	27	55	4	1	324	—	—
13	26	55	5	1	—	4.67	58.35
13	25	55	6	1	309	—	—
13	22	55	9	1	300	—	—

TABLE 3
The Properties of Chalcogenide Glasses Containing Iodine

Composition of glasses, at.%					T_{ds}	ρ	
Sb	As	Ge	Se	I	(°C)	(g/cm³)	N*10³
—	40	30	20	10	376	4.6173	57.55
—	40	20	30	10	316	4.4334	54.83
—	40	10	40	10	211	4.3998	53.98
—	30	30	30	10	366	4.4686	55.42
—	30	20	40	10	283	4.3647	53.71
—	30	10	50	10	205	4.4046	53.78
—	20	30	40	10	347	4.3708	53.94
—	20	20	50	10	276	4.3204	52.9
—	20	10	60	10	157	4.4165	53.66
—	10	30	50	10	335	4.2738	52.48
—	10	20	60	10	232	4.3275	52.73
—	10	10	70	10	121	4.3588	52.7
10	10	30	40	10	251	4.622	53.2
10	10	10	60	10	150	4.6279	53.92
10	20	30	30	10	313	4.4622	52.3
10	20	10	50	10	200	4.632	49.23
10	30	30	20	10	335	4.7487	51.17
10	30	20	30	10	299	4.7085	55.04
10	—	30	50	10	259	4.5951	53.36
10	—	20	60	10	220	4.521	52.11
10	—	10	70	10	116	4.5538	52.11

TABLE 4

The Properties of Glasses Containing Br and Cl

Composition of glasses at.%						
Sb	As	Ge	S	Br	Cl	T_{ds}
—	20	—	70	10	—	24
—	20	—	60	20	—	–6
—	30	—	40	30	—	–12
—	29	—	29	42	—	–39
—	19	—	34	47	—	–63
—	36.5	—	46	17.5	—	116
—	25	—	65	10	—	68
—	21	—	66.5	12.5	—	31
—	10	10	70	—	10	124
—	20	10	60	—	10	178
—	10	20	60	—	10	242
—	20	20	50	—	10	286
10	20	—	60	—	10	125
—	—	10	80	—	10	122
—	—	20	70	—	10	224
10	—	20	60	—	10	230
10	10	20	50	—	10	274
10	10	30	40	—	10	356

References

1. **N. A. Goryunova, B. T. Kolomiiets** "Semiconductor Properties of Chalcogenide Glasses" *Bull. Izobretenyi Otkryityi* (Moscow) 28, 3, 1971.
2. **N. F. Mott, E. A. Davis** "Electron Processes in Non-Crystalline Materials" Clarendon Press, Oxford, 1979.
3. **Edited by M. H. Brodsky** "Amorphous Semiconductors" Springer-Verlag, New York, 1979.
4. **Z. U. Borisova** "Chemistry of Vitreous Semiconductors" Leningrad State University Press, Leningrad, 1972.
5. **Z. U. Borisova** "Chemistry of Semiconductor Glasses" Leningrad State University Press, Leningrad, 1983.
6. **Edited by R. L. Myuller** "Chemistry of Solid" Leningrad State University Press, Leningrad, 1965.
7. **G. Z. Vinogradova** "Glass-Forming and Phase Equilibriums in Chalcogenide Systems" Nauka, Moscow, 1984.
8. **A. Feltz** "Amorphe und Glasartige Anorganishe Festkorper" Akademie Verlag, Berlin, 1983.
9. **V. S. Minayev** "Vitreous Semiconductor Melts" Metallurgiya, Moscow, 1991.
10. **R. Frerichs** "New Optical Glasses Transparent in the Infrared up to 12 μm" *Phys. Rev.* 78, 643, 1950.
11. **R. Frerichs** "New Optical Glasses with Good Transparency in the Infrared" *J. Opt. Soc. Am.* 43, 1153, 1953.
12. **W. A. Fraser** "A New Triaxial System of Infrared Glasses" *J.O.S.A.* 43, 823, 1953.
13. **G. Dewulf** "Glasses Transparent in the Infrared" *Rev. Opt.* 33, 513, 1954.
14. **F. W. Glaze, D. H. Blackbur, J. S. Osmalov, D. Hubbard, M. H. Black** "Properties of Arsenic Sulphide Glass" *J. Res. Nat. Bur. Stand.* 59, 83, (1957).
15. **J. E. Stanworth** "Glass Formation from Melts of Nonmetallic Compounds of the Type A_xB_y" *Phys. Chem. Glasses* 20, 116, 1979.
16. **N. A. Goryunova, B. T. Kolomiiets** "About the Problem of Glass-Formation Regularities in Chalcogenide Glasses" "Stecloobraznoie Sostoyanije". *Isd. Akad. Nauk SSSR,* Moscow-Leningrad, 1960, 71.
17. **J. C. Philips** "Topology of Covalent Non-Crystalline Solids I" *J. Non-Cryst. Solids* 34, 153, 1979.
18. **A. Dietzel** "Die Kationenfeldstorken und ihre Berziehungen zu Entglasungsvorgangen zur Verbundungsbildung und zu den Schmelzpunkten von Silicaten" *Z. Elektrochem.* 48, 9, 1942.
19. **L. A. Baydakov, L. N. Blinov** "About Correlation between Atomic-Structural Characteristics of Melts and Their Glass-Forming Ability" *Fisika Khim. Stekla* 13, 340, 1987.
20. **S. A. Dembovsky, Ye. A. Chechetkina** "Glass-Formation" Nauka, Moscow, 1990, 23.
21. **B. T. Kolomiiets, B. V. Pavlov** "Optical Properties of Chalcogenide Glasses" "Stekloobraznoie Sostoyanije" *Isd. Akad. Nauk SSSR,* Moscow-Leningrad, 1960, 460.

22. **J. Matsuda** "Studies of Selenium Glass" *Rev. Phys. Chem. Jpn.* 29, 27, 1959.

23. **E. B. Deeg** "Physical Properties of Glasses in the System Arsenic-Sulfur-Halogen" "Advances in Glass Technology Part I" Plenum Press, New York, 1962, 348.

24. **S. S. Flaschen, A. D. Pearson, W. R. Northover** "Formation and Properties of Low-Melting Glasses in the Ternary Systems As–Tl–S, As–Tl–Se and As–Se–S" *J. Am. Ceram. Soc.* 43, 274, 1960.

25. **N. A. Goryunova, B. T. Kolomiiets** "New Vitreous Semiconductors" *Izv. Akad. Nauk SSSR* Ser. Fiz. 20, 1496, 1956.

26. **N. A. Goryunova, B. T. Kolomiiets, V. P. Shilo** "Glass-Formation Region in the Melts of Arsenic, Thallium and Antimony Chalcogenides" *Zh. Tekh. Fiz.* 28, 981, 1958.

27. **B. T. Kolomiiets, T. N. Mamontova, T. Ph. Nazarova** "Electrical Properties of Chalcogenide Glasses" Stekloobraznoie Sostoyanije, *Isd. Akad. Nauk SSSR.* Moscow-Leningrad, 1960, 465.

28. **V. A. Khar'yuzov, K. S. Yevstrop'yev** "Electrical Conductivity of Vitreous Melts in the System Selenium-Germanium" *OMP* 10, 17, 1961.

29. **Z. U. Borisova, Chen-Sai, R. L. Myuller** "About Electrical Conductivity of Vitreous GeSe$_x$" *Zh. Prik. Khim.* 35, 755, 1962.

30. **J. Lecomte** "Le Rayonnement Infrarouge t.I" Gauthier-Villars, editeur, Paris 1948, 223.

31. **E. Brandenberger, V. Epprecht** "Rontgenographische Chemie" Birkhauser Verlag, Basel, 1963, 47.

32. **A. F. Wells** "Structural Inorganic Chemistry" Clarendon Press, Oxford, 1945, 123.

33. **P. P. Kobeko** "Amorphous Substances" *Isd. Akad. Nauk SSSR,* Moscow, 1952, 50.

34. **O. V. Mazurin, Ye. A. Porai-Koshits** "About Some Moot Points in Glass Terminology" *Steklo Keram.* 12, 13, 1975.

35. **G. Tammann** "Glasses as Supercold Liquids" *J. Soc. Glass Technol.* 9, 166, 1925.

36. **A. A. Lebedev** "About Polymorphism and Glass Annealing" *Tr. GOI* 1, 1921.

37. **A. A. Lebedev** "Investigation of Glass Structures by Spectral-Optical Methods" "Stekloobraznoie Sostoyanije," *Isd. Akad. Nauk SSSR,* Moscow-Leningrad, 1960, 7.

38. **J. T. Randall, H. P. Rooksby, B. S. Cooper** "The Structure of Glasses; the Evidence of X-Ray Diffraction" *J. Soc. Glass Technol.* 14, 219, 1930.

39. **N. N. Valenkov, Ye. A. Porai-Koshits** "X-Ray Investigations of Glasses in the System Na_2O-SiO_2 Phisiko-Khemicheskiye Svoistva Troinoi Sistemy Na_2O-PbO-SiO_2" *Isd. Akad. Nauk SSSR,* Moscow-Leningrad, 1949, 147.

40. **W. H. Zachariasen** "The Atomic Arrangement in Glasses" *J. Am. Chem. Soc.* 54, 3841, 1932.

41. **B. E. Warren** "X-Ray Determination of the Structure of Glass" *J. Am. Ceram. Soc.* 17, 249, 1934.

42. **G. Hägg** "The Vitreous State" *J. Chem. Phys.* 3, 42, 1935.

43. **W. H. Zachariasen** "The Vitreous State" *J. Chem. Phys.* 3, 162, 1935.

44. **R. L. Myuller** "Hard Glasses' Structure by the Electric Conductivity Evidence" *Izv. Akad. Nauk SSSR* Ser. Phiz. 4, 607, 1940.

45. **K. H. Sun** "Fundamental Condition of Glass Formation" *J. Am. Ceram. Soc.* 30, 277, 1947.

46. **H. Rawson** "Inorganic Glass-Formed Systems" Academic Press, New York, 1967, 257.

47. **Ya. K. Syrkin, M. Ye. Dyatkina** "Chemical Bond and Molecular Structure" Goskhimizdat, Moscow-Leningrad, 1946, 406.

48. **R. L. Myuller** "Electric Conductivity of Hard Ion-Atom-Valent Substances" *Zh. Tekh. Fiz.* 26, 2614, 1956.

49. **A. Smekal** "On the Structure of Glass" *J. Soc. Glass Technol.* 35, 411, 1951.

50. **I. E. Stanworth** "The Structure of Glass" *J. Soc. Glass Technol.* 30, 54, 1946.

51. **A. Winter-Klein** "Les Formateurs des Verres et le Tableau Periodique des Elements" *Verres Refract.* 9, 147, 1955.

52. **W. Hume-Rothery** "The Crystal Structure of Elements of the B-Subgroups and Their Connection with the Periodic Table and Atomic Structures" *Philos. Mag.* 7–9, 65, 1930.

53. **C. A. Coulson** "Valence" Clarendon Press, Oxford, 1952, 265.

54. **N. F. Mott** "Conduction in Glasses Containing Transition Metal Ions" *J. Non-Cryst. Solids* 1, 1, 1968.

55. **T. A. Sidorov, N. N. Sobolev, Ye. P. Markin, V. V. Obukhov-Denisov, V. P. Cheremsinov** "Oscillation Spectra and Structure of Glass-Forming Oxides in Crystalline and Vitreous State" "Stekloobraznoie Sostoyanije" *Isd. Akad. Nauk SSSR,* Moscow-Leningrad, 1960, 207.

56. **W. A. Weyl** "Colored Glasses" Sheffeld, 1951.

57. **J. M. Stewels** "Glass Considered as a Polymer" *Glass Ind.* 35, 135, 1954.

58. **V. V. Tarasov** "Glass Considered as a Polymer" "Stekloobraznoie Sostoyanije", *Isd. Akad. Nauk SSSR,* Moscow-Leningrad, 1960, 78.

59. **J. C. Phillips** "Realization of Zachariasen Glass" *Solid State Commun.* 47, 203, 1983.

60. **R. Hoseman, M. P. Hentschel, V. Schmeisser, R. Bruckner** "Structural Model of Vitreous Silica Based on Microparacrystal Principles" *J. Non-Cryst. Solids* 83, 223, 1986.

61. **G. Z. Pinsker** "Determination of Lattice Regularities in Amorphous Structure" *Phis. Khim. Stekla* 6, 521, 1980.

62. **J. R. Rollo, G. R. Burns** "An Inorganic Molecular Glass; a Glass Transition for Molecular Tetraphosphorous Triselenide" *J. Non-Cryst. Solids* 127, 242, 1991.

63. **G. Lucovsky, R. M. Martin** "A Molecular Model for the Vibrational Modes in Chalcogenide Glasses" *J. Non-Cryst. Solids* 8–10, 185, 1972.

64. **H. Krebs** "Anorganische Hochpolimere" *Angew. Chem.* 70, 615, 1958.

65. **R. L. Myuller** "Chemical Peculiarities of Polymer Glass-Forming Substances and Glass-Formation Nature" "Stekloobraznoie Sostoyanije" *Isd. Akad. Nauk SSSR,* Moscow-Leningrad, 1960, 61.

66. **V. V. Tarasov** "Polymer Models and Properties of Boric Anhydride and Boric Glasses" "Stekloobraznoie Sostoyanije" *Isd. Akad. Nauk SSSR,* Moscow-Leningrad, 1965, 28.

67. **V. V. Tarasov** "On Polymer Character of Vitreous Arsenic Sulphide" "Stekloobraznoie Sostoyanije" *Isd. Akad. Nauk SSSR,* Moscow-Leningrad, 1965, 167.

68. **L. Pauling** "Die Natur der Chemischen Bindung" Verlag Chemie, Weinheim, 1962, 94.

69. **V. A. Funtikov** "Structural Peculiarities of Chalcogenide Glasses" *Fis. Khim. Stekla* 19, 226, 1993.

70. **S. R. Elliott** "Medium-Range Order in Amorphous Materials: Documented Cases" *J. Non-Cryst. Solids* 97–98, 159, 1987.

71. **D. L. Price, S. Susman, A. C. Wright** "Probing Medium-Range Order in Chalcogenide Glasses by Neutron Scattering and Optical Spectroscopy" *J. Non-Cryst. Solids* 97–98, 167, 1987.

72. **C. S. Barret** "Structure of Metals" Mettalurgiya, Moscow, 1948, 290.

73. **K. S. Yevstrop'yev** "Connection Between Structure and Properties of Glasses" "Stekloobraznoie Sostoyanije" *Isd. Akad. Nauk SSSR,* Moscow-Leningrad, 1960, 39.

74. **S. V. Nemilov** "On Interrelation of Free Activation Energy of Viscous Flow and Energy of Chemical Bond in Glasses" *Fyizika Tvyordogo Tyela,* 6, 1375, 1964.

75. "Metody Ispytanii na Mikrotvyordost'. Pryibory" Nauka, Moscow, 1965.

76. **S. Ye. Sorkin** "Quartz Vertical Dilatometer QVD" *Isd. Gos. Inst. Stekla,* Moscow, 1964.

77. **V. T. Slavyanski, Ye. N. Krestnikova, V. M. Boreiko** "The Method for Glass Viscosity Determination" *Steklo Keram.* 4, 18, 1962.

78. **S. V. Nemilov** I. "Viscosity and Structure of Glasses in the System AsGeSe in the Region with High Selenium Content" *Zh. Prikl. Khyim.* 37 1452 1964; II. "Viscosity and Structure of Glasses in the System As–Ge–Se in the Region of Low Selenium Content" *Zh. Prikl. Khyim.* 37, 1699, 1964.

79. **A. R. Hilton, C. E. Jones, M. Brau** Part 1. "New High Temperature Infrared Transmitting Glasses" *Infrared Phys.* 4 213 1964; Part 2. "New High Temperature Infrared Transmitting Glasses" *Infrared Phys.* 6, 183, 1966.

80. **A. R. Hilton, C. E. Jones, M. Brau** "Nonoxide Chalcogenide Glasses. Glass-Forming Region and Variation in Physical Properties" *Phys. Chem. Glasses* 7, 105, 1966.

81. **J. A. Savage, S. Nielsen** "The Infrared Transmission of Telluride Glasses" *Phys. Chem. Glasses* 7, 56, 1966.

82. **B. T. Kolomiiets, N. A. Goryunova, V. P. Shilo** "Vitreous State in Chalcogenides" "Stekloobraznoie Sostoyanije" *Isd. Akad. Nauk SSSR,* Moscow-Leningrad, 1960, 456.

83. **L. A. Baydakov** "Electric Conductivity in the Vitreous System As–Ge–Se" Vestn. LGU 22, 105, 1962.

84. **J. A. Savage, S. Nielsen** "Chalcogenide Glasses Transmitting in the Infrared between 1–20 microns" *Infrared Phys.* 6, 195, 1965.

85. **R. W. Haisty, H. Krebs** "Electric Conductivity of Melts and Their Ability to Form Glasses. II. The Ge–As–Se System" *J. Non-Cryst. Solids* 1, 427, 1969.

86. **P. J. Webber, J. A. Savage** "Some Physical Properties of Ge–As–Se Infrared Optical Glasses" *J. Non-Cryst. Solids* 20, 271, 1976.

87. **Z. U. Borisova, A. V. Pasin** "Investigation of the Vitreous System Antimony-Germanium-Selenium" *"Khimija Tvyordogo Tyela"* Isd. LGU 1965, 98.

88. **R. W. Haisty, H. Krebs** "Electric Conductivity of Melts and Their Ability to Form Glasses. I. The Ge–Sb–Se System" *J. Non-Cryst. Solids* 1, 399, 1969.

89. **M. Brau, R. E. Johnson, R. J. Patterson** "Ge–Sb–Se Glasses Compositions" U.S. Patent No. 3360649, 1967.

90. **D. Linke, F. Heyder** "Das System Ge–Sb–Se" *J. Anorg. Allgem. Chem.* 425, 155, 1976.

91. **A. V. Pasin, G. M. Orlova** "Crystallization of the Glasses $SbGe_xSe_y$" *Zh. Prikl. Khim.* 47, 505, 1974.

92. **A. V. Pasin, Z. U. Borisova** "Glass-Formation Region in the System Bi–Ge–Se" *Vestn. LGU Ser. Phyz. Khim.* 4, 140, 1969.

93. **A. V. Pasin, Z. U. Borisova, M. A. Garbuzova** "Investigation of Glasses in the System Bi–Ge–Se in the Critical Temperature Region" *Neorg. Matyer.* 6, 884, 1970.

94. **J. C. Malaurent, J. I. Dixmier** "X-Ray Diffraction and Local Order Modelling of Ge_xSe_{1-x} Amorphous Alloys" *J. Non-Cryst. Solids* 35–36, 1227, 1980.

95. **J. N. Bryden** "The Crystal Structures of the Germanium-Arsenic Compounds. I. Germanium Diarsenid, $GeAs_2$" *Acta Crystal.* 15, 167, 1962.

96. **A. V. Pasin, Z. U. Borisova** "Electric Conductivity of the Glasses in the System Sb–Ge–Se" *Neorg. Matyer.* 5, 1903, 1969.

97. **J. A. Savage, P. I. Webber, A. M. Pitt** "An Assessment of Ge–Sb–Se Glasses as 8 to 12 μm Infrared Optical Materials" *J. Mater. Sci.* 13, 859, 1978.

98. **G. M. Orlova, I. I. Kozhyina, V. G. Korolenko** "Structural Diagram of the System Sb_2Se_3–$GeSe_2$, Sb_2Se_3–$GeSe$" *Vestn. LGU Ser. Phyz. Khyim.* 4, 90, 1973.

99. **N. A. Korepanova, G. A. Chalabyan, G. M. Orlova** "Thermal Expansion and Ultrasound Spreading Speed in the Glasses of the System Sb–Ge–Se" *Zh. Prikl. Khyim.* 49, 1173, 1976.

100. **M. S. Gutenev, I. V. Viktorovsky, L. A. Baydakov, A. V. Pasin** "Peculiarities of Structure-Chemical Structure of the Glasses in the System Sb–Ge–Se According to the Evidence from the Investigation of Dielectrical and Magnetic Properties" *Phyz. Khym. Stekla* 1, 350, 1975.

101. **F. M. Gashimzade, V. Ye. Khartzijev** "Structures SnS, GeS, GeSe" *Fyiz. Tvyordogo Tyela* 4, 434, 1962.

102. **H. Krebs** "Mischkristalle der Chalkogenide IV Gruppe" *Z. Anorg. Allg. Chem.* 1–2, 334, 1964.

103. **Ye. V. Shkol'nikov, Z. U. Borisova** "Electric Conductivity and Microhardness of Vitreous and Glass-Crystalline $AsSe_{1.5}Pb_x$," Vestn. LGU 4, 120, 1965.

104. **J. J. Petz, R. F. Kruh, G. C. Amstutz** "X-Ray Diffraction Study of Lead Sulfide-Arsenic Sulfide Glasses" *J. Chem. Phys.* 34, 526, 1961.

105. **J. A. Savage** "Optical Properties of Chalcogenide Glasses" *J. Non-Cryst. Solids* 47, 101, 1982.

106. **J. Jerger** "Glass Composition" U.S. Patents 2.883, 292–295; *J. Am. Ceram. Soc.* 42 P(9), 234df, 1959.

107. **A. A. Obraztsov, G. M. Orlova** "Crystallization of Glasses in the System As–Se–Te" *Neorg. Mater.* 7, 2166, 1971.

108. **Z. G. Ivanova, V. S. Vasyiliev, K. G. Vasyilieva** "New IR-Transmitting Chalcohalide Glasses" *J. Non-Cryst. Solids* 162, 123, 1993.

109. **S. A. Dembovsky** "Glass-Formation in the System Ge–Tl–Se" *Zh. Neorg. Khyim.* 13, 1721, 1968.

110. **Ye. Yu. Turkina, N. N. Koshina, G. M. Orlova, A. A. Obraztsov** "Components' Interrelation in the Ternary System Tl–Ge–Se along the Cuts $Tl_2Se–GeSe_2$ and TlSe–Ge" *Zh. Neorg. Khim.* 23, 497, 1978.

111. **D. Linke, M. Ditter, F. Krug** "Glasbildung und Phasentrennung in den System Tl–Ge–Se und Pb–Ge–Se" *Z. Anorg. Allg. Chem.* 444, 217, 1978.

112. **A. Feltz, F. J. Lippmann** "Zur Glasbildung in System Ge–Se" *Z. Anorg. Allg. Chem.* 398, 157, 1973.

113. **L. Cervinski, O. Smotlacha, J. Bergerova, L. Tichy** "The Structure of Glassy Ge–Sb–S System and Its Connection with the MRO Structures of GeS_2 and Sb_2S_3," *J. Non-Cryst. Solids* 137–138 (1), 123, 1991.

114. **S. S. Flaschen, A. D. Pearson, W. R. Northover** "Low Melting Sulfide-Halogen Inorganic Glasses" *J. Appl. Phys.* 31, 219, 1960.

115. **F. G. Lin, S. N. Ho** "Chemical Durability of Arsenic-Sulfur-Iodine Glasses" *J. Am. Ceram. Soc.* 46, 24, 1963.

116. **T. E. Hopkins, R. A. Pasternak, F. S. Gould, J. R. Herndon** "X-Ray Diffraction Study of Arsenic-Trisulfide-Iodine Glasses" *Phys. Chem.* 66, 733, 1962.

117. **Z. U. Borisova, L. N. Doynikov** "Investigation of Glassy Melts $AsSe_xI_y$" Stekloobraznoie Sostoyanije *Isd. Akad. Nauk SSSR,* Moscow-Leningrad, 1965, 181.

118. **A. D. Pearson, W. R. Northover, I. F. Dewald, W. F. Peck** "Chemical, Physical and Electrical Properties of Some Unusual Inorganic Glasses" Advances in Glass Technology, Part 1, Plenum Press, New York, 1962, 357.

119. **D. L. Eaton** "Electrical Conduction Anomaly of Semi-Conducting Glasses in the System As–Te–I" *J. Am. Ceram. Soc.* 47, 554, 1964.

120. **A. G. Fischer, A. S. Mason** "Properties of an As–S–Br Glasses" *J. Opt. Soc. Am.* 52, 721, 1962.

121. **E. W. Deeg, G. M. Habaschi, R. O. Loutfi** "Preparation, Properties and Structure of As–S–Cl Glasses" "Herstellung, Eingenscaften und Struktur der Glaser im System As–S–Cl" Sprechsaal fur Keramikglass email Silicate I. 100, 757, 1967; II. 100, 793, 1967; III. 100, 873, 1967.

122. **S. A. Dembovsky, N. P. Popova** "Glass-Formation in the Systems Ge–Se–I and Si–Se–I" *Neorg. Mater.* 6, 138, 1970.

123. **A. P. Chernov, S. A. Dembovsky, N. P. Luzhnaya** "Investigation of the Ternary System As–Te–I" *Zh. Neorg. Khyim.* 20, 2174, 1975.

124. **A. P. Chernov, S. A. Dembovsky, V. I. Makhova** "Viscosity and Glass Structure in the Systems As_2X_3–AsI_3," *Neorg. Mater.* 6, 823, 1970.

125. **V. V. Kirilenko, S. A. Dembovsky** "Glass-Formation and Chemical Compounds in the Systems A^{IV}-B^{VI}-C^{VII}(A^{IV}-Si, Ge, B^{VI}-S, Se, C^{VII}-Br, I)" *Phyz. Khim. Stekla* 1 225, 1975.

126. **A. B. Pearson** "Preparation and Properties of Sulfide, Selenide and Telluride Glasses" *Glass Ind.* 46, 18, 1965.

127. **R. F. Johnson, R. J. Patterson, A. E. Tilton** "Amorphous Glass Composition" U.S. Patent 3.505.522, cl.250–83, 1970.

128. **K. Quin-Rod, R. F. Johnson** "Compositional Dependence of the Thermal, Structural and Electrical Properties of As–Te–I Glasses" *J. Non-Cryst. Solids* 7, 53, 1972.

129. **O. V. Khyiminetz, V. S. Gerasimenko, V. V. Khyiminetz, I. D. Turyanitza** "Glass-Formation in the System Sb–Se–Br" *Elektron. Tekh. Mater.* 7, 191, 1977.

130. **A. R. Hilton, M. Brau** "New High Temperature Infrared Transmitting Glasses" *Infrared Phys.* 3, 69, 1963.

131. **J. A. Savage, S. Nielsen** "Preparation of Glasses Transmitting in the Infrared between 8 and 15 microns" *Phys. Chem. Glass* 5, 82, 1964.

132. **B. T. Kolomiietz, V. N. Shilo** "Softening Temperature of Some Chalcogenide Glasses" *Steklo Keram.* 8, 10, 1963.

133. **S. V. Nemilov, G. T. Petrovsky** "New Method of Glass Viscosity Measuring" *Zh. Prikl. Khim.* 36, 222, 1963.

134. **S. V. Nemilov** "Viscosity and Structure of Glasses Ge–Se" *Zh. Prikl. Khim.* 37, 1020, 1964.

135. **V. I. Vyedeneiev, L. V. Gurvich, V. N. Kondrat'ev, V. A. Medvedev, Ye. L. Franievich** "Energies of Chemical Bonds' Breakages" *Isd. Akad. Nauk SSSR,* Moscow-Leningrad, 1962.

136. **K. Tanaka** "Structural Phase Transition in Chalcogenide Glasses" *Phys. Rev.* B 39, 1270, 1989.

137. **A. Giridhar, S. Mahadevan** "The Tg Versus Z Dependence of Glasses of the Ge–In–Se System" *J. Non-Cryst. Solids* 151, 245, 1992.

138. **D. I. Tziulyanu, N. A. Gumenyuk** "Structure-Chemical Peculiarities and Optical Properties of Glasses, Enriched with Sulfur, in the System As–S–Ge" *Neorg. Mater.* 29, 689, 1993.

139. **A. N. Sreeram, A. K. Varshneya, D. R. Swiler** I. "Microhardness and Indentation Toughness Versus Average Coordination Number in Isostructural Chalcogenide Glass Systems" *J. Non-Cryst. Solids* 130 225 1991; II. "Gibbs-DiMarzio Equation to Describe the Glass Transition Temperature Trends in Multicomponent Chalcogenide Glasses" *J. Non-Cryst. Solids* 127, 287, 1991.

140. **H. A. Gebbie, C. G. Cannon** "Properties of Amorphous Selenium and Its Use as Optical Material" *J.O.S.A.* 42, 277, 1952.

141. **S. S. Ballard, McCarthy, W. I. Wolfe** "Optical Materials for IR Instrumentation" IRIA University of Michigan State of the Art Report 1959.

142. **A. Vashko** "On the Absorption and Reflection Spectrum of Amorphous Selenium in the Infrared Region" *Czech. J. Phys.* 15, 170, 1965.

143. **A. Vashko, G. Prokopova, B. T. Kolomiiets, B. V. Pavlov, V. N. Shilo** "Absorption Spectra of the Glasses in the System As_2S_3–As_2Se_3" *Opt. Spektrosk.* 12, 275, 1962.

144. **J. T. Edmond, M. W. Redfearn** "Infrared Study of As_2Se_3-type Glasses in the Wavelength Range 1.25 to 25 μm" *Proc. Phys. Soc.* 81, 380, 1963.

145. **J. A. Savage, S. Nielsen** "The Effect of Oxygen on the Infrared Transmission of Ge–As–Se Glasses" *Phys. Chem. Glasses* 6, 90, 1965.

146. **J. A. Savage, P. Q. Webber, A. H. Pitt** "The Potential of Ge–Se–As–Te Glasses as 2–5 μm and 8–12 μm Infrared Optical Materials" *Infrared Phys.* 20, 313, 1980.

147. **T. Kanamori, V. Terunuma, S. Takanashi, T. Miyashita** "Chalcogenide Glass Fibres for Mid-Infrared Transmission" *J. Lightwave Technol.* IT-2, 607, 1984.

148. **T. Kanamori, V. Terunuma** "Transmission Loss Characteristics of $As_{40}Se_{60}$ and $As_{38}Ge_5Se_{57}$ Glass Unclad Fibres" *J. Non-Cryst. Solids* 69, 231, 1985.

149. **V. G. Borisevich, V. G. Plotnichenko, I. V. Skripachev, M. F. Chyurbanov** "Extinction Coefficient of the Groups S–H in Vitreous Arsenic Sulphide" *Vysokochist. Vesch.* 4, 198, 1990.

150. **I. V. Skripachev, T. G. Devyatykh, M. F. Churbanov, V. A. Boyko, A. M. Bagrov** "Highly Pure Chalcogenide Glasses for Fibre Optics" *Vysokochist. Vesch.* 1, 121, 1987.

151. **I. V. Skripachov, T. G. Devyatykh, M. F. Churbanov, V. A. Shipunov** "Determination of the Extinction Coefficient of the Groups Se–H in Vitreous Selenium" *Vysokochist. Vesch.* 2, 91, 1989.

152. **T. I. Bychkova, G. Z. Vinogradova, V. V. Voytzekhovsky, V. G. Plotnichenko** "Intrinsic Absorption As_2O_3 in Vitreous Selenium" *Vysokochist. Vesch.* 4, 203, 1990.

153. **C. T. Moynihan, P. B. Macedo, M. S. Maklad, R. K. Mohr, R. E. Howard** "Intrinsic and Impurity Infrared Absorption in As_2Se_3 Glass" *J. Non-Cryst. Solids* 17, 369, 1975.

154. **D. Lezal, I. Srb** "Preparation of High Purity Chalcogenide Glasses" 11th Int. Congr. Glasses (Prague) 5, 497, 1977.

155. **D. Lezal, I. Srb** "Synthesis of Chalcogenide Glasses of Se–Ge and Se–As Systems without Traces of Oxygen" *Collect. Czech. Chem.* 36, 2091, 1971.

156. **V. S. Gerasimenko, D. I. Bletzkan, M. Yu. Sichka** "IR-Transmission Spectra and Structure of Glasses $GeSe_x$" *Ukr. Phys. Zh.* 21, 1481, 1990.

157. **T. Ohsaka** "Infrared Studies of Se-Based Polymary Chalcogenide Glasses (II): $Y_XZ_XSe_{100-2X}$ (Y = Ge, As; Z = S, Te)" *J. Non-Cryst. Solids* 22, 89, 1976.

158. **A. R. Hilton, C. E. Jones** "Non-Oxide IVA-VIA-VA Chalcogenide Glasses. II. Infrared Absorption by Oxide Impurities" *Phys. Chem. Glass* 7, 112, 1966.

159. **I. V. Skripachev, V. V. Kuznetsov, V. G. Plotnichenko, A. A. Pushkin, M. F. Churbanov, V. A. Shipunov** "Investigation of Extrinsic Composition of the Glasses in the System Ge–S, Obtained by Direct Synthesis from Elementary Substances" *Vysokochist. Vesch.* 6, 208, 1987.

160. **A. R. Hilton, C. E. Jones** "The Thermal Change in the Non-Dispersive Infrared Refractive Index of Optical Materials" *Appl. Opt.* 6, 1513, 1967.

161. **P. W. Kruse, L. D. McClauchlin, R. B. McQuistan** "Elements of Infrared Technology" John Wiley & Sons, New York, 1962, 164.

162. **Ye. Ja. Mukhin, N. G. Gutkina** "Crystallization of Glasses and Methods of Its Prevention" OBORONGIZ, Moscow, 1960.

163. **G. Z. Vinogradova, S. A. Dembovsky, N. P. Luzhnaya** "Crystallization Fields of the Glasses As–Ge–Se in the Glass-Formation Region" *Zh. Neorg. Khym.* 13, 1444, 1968.

164. **M. P. Susarev, N. S. Martyinova** "Unified Calculation Method for Compositions of Ternary Eutectics and Azeotrops by Binary Data" *Zh. Prikl. Khim.* 52, 556, 1979.

165. **I. M. Buzhinsky, L. I. Dyomkina** "Properties of Glasses, Used at Optical Systems Construction" Collection Physiko-Khimicheskie Osnovy Proizvodstva Opticheskogo Stekla, Isd. Khimija, Leningrad, 1976, 6.

166. **N. A. Korepanova, G. M. Orlova, A. V. Pasin** "Investigation of Structure-Chemical Peculiarities of the Glasses in the System Sb–Ge–Se by Viscosimetry" *Zh. Prikl. Khim.* 49, 36, 1976.

167. **N. Kh. Abrikosov, Ye. S. Avilov, O. G. Karpinski, O. V. Radkevich, A. Ye. Shalimova** "Phase Transfers in the Melts $Ge_{1-Y}(Te_{1-X}Se_X)_Y$" *Isv. Akad. Nauk SSSR Neorg. Mater.* 22, 33, 1984.

168. **N. Kh. Abrikosov, Ye. S. Avilov, O. G. Karpinski, A. Ye. Shalimova** "Some Peculiarities of Hard Solutions Based on Germanium Telluride" *Neorg. Mater.* 16, 237, 1980.

169. **T. I. Darvoid, V. I. Kovalev, I. S. Lisitsky, V. S. Mironov, F. S. Faizullov** "Investigation of the Crystals KRS5 and KRS6 Resistance to the Radiation of Impulse CO_2-Laser" *Kvant. Electron.* 1, 2172, 1974.

170. **F. Horrigan, C. Klein, R. Rudko, D. Wilson** "Windows for Highpower Lasers" *Microwaves Laser Technol. Soc.* 1, 68, 1969.

171. **T. F. Deutsch** "Absorption Coefficient of Infrared Laser Window Materials" *J. Phys. Chem. Soc.* 34, 2091, 1973.

172. **M. Sparks** "Stress and Temperature Analysis for Surface Cooling or Heating of Laser Window Materials" *J. Appl. Phys.* 44, 4137, 1973.

173. **M. D. Rechtin, A. R. Hilton, D. J. Hayes** "Infrared Transmission in Ge–Sb–Se Glasses" *J. Electron. Mater.* 4, 347, 1975.

174. **V. G. Artyushenko, Ye. M. Dianov, Ye. P. Nikitin** "Calorimetric Method for Determination of Volume and Surface Absorption in the Materials, Transparent in IR-Range" *Kvant. Electron.* 5, 1065, 1978.

175. **V. I. Kovalev, V. V. Morozov, F. S. Faizullov** "Opacity Appearance and Optical Materials Destruction under the Impulse Laser Attack on the Carbon Dioxide" *Kvant. Elektron.* 1, 2172, 1974.

176. **A. R. Hilton, D. J. Hayes, M. D. Rechtin** "Infrared Absorption of Some High-Purity Chalcogenide Glasses" *J. Non-Cryst. Solids* 17, 319, 1975.

177. **D. Lezal, I. Srb** "Purification of As_2Se_3 by Decomposition of Urea" *Collect. Czech. Chem. Commun.* 36, 3732, 1971.

178. **M. Brau, R. Ye. Johnson, R. J. Patterson** "Ge–Sb–Se Glass Compositions" U.S. Patent No. 3360649, 1967.

179. **I. Srb, Z. Wachtl** Chechoslovakija Patent No. 133171, 1969.

180. **A. M. Reitter, A. N. Sreeram, A. K. Varshneva, D. R. Swiler** "Modified Preparation Procedure for Laboratory Melting of Multicomponent Chalcogenide Glasses" *J. Non-Cryst. Solids* 139, 121, 1992.

181. **R. A. Hilton, G. R. Cronin** "Production of Infrared Optical Materials at Amorphous Materials Incorporated" SPIE 618 *Infrared Opt. Mater. and Fiber* 4 184, 1986,

182. **S. P. Vikhrov, V. N. Ampilogov, V. S. Minayev** "Investigation of the Influence of Synthesis Conditions and Composition upon the Oxygen and Carbon Content in Vitreous Chalcogenide Semiconductors" *Phys. Khim. Stekla* 10, 486, 1984.

183. **G. G. Devyatikh, M. F. Churbanov** "Determination of Microimpurities in Chalcogens" *Vysokochist. Vesch.* 1, 32, 1990.

184. **Ye. M. Dianov, M. Yu. Petrov, V. G. Plotnichenko, V. K. Sysoiev** "Evaluation of Minimum Optical Losses in Chalcogenide Glasses" *Kvant. Electron.* 9, 793, 1982.

185. **A. M. Bagrov, P. I. Baikalov, A. V. Vasyliev, G. G. Devyatikh, Ye. M. Dianov, V. G. Plotnichenko** "Fibre Lightguides of Middle Infrared Range Based on As–S and As–Se with Optical Losses Less than $1^{dB}/m$" *Kvant. Electron.* 10, 1906, 1983.

186. **G. G. Devyatikh, Ye. M. Dianov, B. G. Plotnichenko, I. V. Skripachev, M. F. Churbanov, V. A. Shipunov** "Heterophase Extrinsic Insertions in Fibre Lightguides from Chalcogenide Glasses" *Vysokochist. Vesch.* 4, 192, 1990.

187. **V. G. Borisevich, V. V. Voytzekhovski, I. V. Skryipachov, V. G. Plotnichenko, M. F. Churbanov** "Investigation of the Influence of Extrinsic Hydrogen on the Optical Properties of Chalcogenide Glasses in the System As–Se" *Vysokochist. Vesch.* 1, 65, 1991.

188. **V. G. Borisevich, V. G. Plotnichenko, I. V. Skripachov, G. Ye. Snopatin** "Selective Absorption Bands, Caused by Carbon Compounds in the Transmission Spectra of the Glasses As–S and As–Se" *Vysokochist. Vesch.* 5, 110, 1994.

189. **T. Katsuyama, K. Ishida, S. Satoh, H. Matsumura** "Low Loss Ge–Se Chalcogenide Glass Optical Fibres" *Appl. Phys. Lett.* 45, 925, 1984.

190. **I. P. Parant, G. Sergent, D. Guignot, C. Brehu** "Chalcogenide Glass Optical Fibres" *Glass Technol.* 24, 161, 1983.

191. **Yu. A. Chulzhanov** "High-Pure Tellurium Production" IV. Vsesoyusnoje Soveschanije "Khimija i Technologyia Khalcogenov i Khalcogenidov", *Karaganda,* 1990, 12.

192. **V. A. Pazukhin, A. Ya. Fisher** "Separation and Refining of Metals in Vacuum" Isd. Mettalurgija, Moscow, 1969, 134.

193. **T. D. Portnaya, S. P. Byilam** "Investigation of Possibility of Commercial As_2S_3 Purification of Oxide Impurities with the Help of Gas Absorber" *Zh. Prikl. Khim.* 54, 2691, 1981.

194. **A. A. Godovikov, V. G. Nenashev, A. P. Andreev** "On As_2S_3 Decomposition at Sublimation in Vacuum" *Dokl. Akad. Nauk SSSR* 212, 1196, 1973.

195. **D. S. Cary, W. F. Parsons, E. Carnall** "Infrared Lens Cement" U.S. Patent 3.157.521 cl. 106–47, 1, 1964.

196. **Ye. I. Karison, J. V. Kirhan** "Immersed Bolometers and Immersion Glasses Therefore" U.S. Patent 3.121.203 cl. 338–18, 1, 1964.

197. **A. G. Fischer, C. J. Nuese** "Highly Refractive Glasses to Improve Electroluminescent Diode Efficiency" *J. Electrochem. Soc.* 116, 1718, 1969.

198. **I. D. Turyanitza, V. V. Khiminetz, O. V. Khiminetz** "Investigation of Interreaction and Glass-Formation in Ternary Systems As–S(Se, Te)–Br" *Phys. Khim. Stekla* 1, 190, 1975.

199. **Yu. S. Akimov, V. V. Garshenin, N. I. Dovgashey, I. D. Turyanitza, S. A. Charykov, D. V. Chepur** "Chalcogenide Glasses and Their Application in Optical Electronics" Isd. CNII Electronika, Moscow, 1973.

200. **G. V. Saidov, M. Ye. Yudovich** "Liquid Optical Element BAIR with Variable Number of Insertions" *Opt. Spectrosk.* 36, 1216, 1974.

201. **V. M. Zolotarev** "BAIR Attachment and Determination of Optical Constants on Standard IR-Spectrophotometers" *OMP* 8, 45, 1976.

202. **R. L. Abrams, D. A. Pinnow** "Acousto-Optic Properties of Crystalline Germanium" *J. Appl. Phys.* 41, 2765, 1970.

203. **J. T. Krause, C. R. Kurkjian, D. A. Pinnow, E. A. Sigety** "Low Acoustic Loss Chalcogenide Glasses — a New Category of Materials for Acoustic and Acousto-Optic Application" *Appl. Phys. Lett.* 17, 367, 1970.

204. **V. Ohmachi, N. Ishida** "Vitreous As_2Se_3, Investigation, Acousto-Optical Properties and Applications to Infrared Modulators and Scanners" *J. Appl. Phys.* 43, 1709, 1972.

205. **A. W. Warner, D. A. Pinnow** "Miniature Acousto-Optic Modulator for Optical Communications" *J.E.E.E.J. Quantum. Electron.* QE-9 12, 1153, 1973.

206. **D. V. Sheloput, V. F. Glushkov** "Acoustic Characteristics of Chalcogenide Glasses" *Isv. Akad. Nauk SSSR Serija Neorg. Mater.* 9, 1149, 1973.

207. **R. W. Dixon, M. G. Cohen** "A New Technique for Measuring Magnitudes of Photoelastic Tensors and Its Application to Lithium Niobate" *Appl. Phys. Lett.* 8, 205, 1966.

208. **Ye. E. Danyushevsky** "Principles of Linear Annealing of the Optical Glass" OBO-RONGIZ, Moscow, 1959.

209. **L. I. Dyomkina** Collection "Physiko-Khimicheskije Osnovy Proizvodstva Opticheskogo Stekla" Isd. Khymija, Leningrad, 1976.

210. **V. S. Doladugina, V. I. Voskresenskaya, Yu. D. Pushkin** "Michelson Infrared Interferometer" *OMP* 4, 37, 1973.

211. **V. S. Doladugina** "Experimental Determination of the Sensitivity of Luminous Point Method" *OMP* 10, 12, 1972.

212. **M. S. Gomelsky** "Fine Annealing of Optical Glass" Isd. Mashinostroyenije, Moscow, 1969.

213. **Z. I. Kanchiev** "Influence of Structural Changes in Chalcogenide Glasses on the Character of Temperature Expansion" *Phys. Khim. Stekla* 1, 547, 1975.

214. **A. S. Pashinkin, A. S. Malkova, A. D. Chervonny** "Thermodynamics of the Evaporation As_2S_3" *Neorg. Mater.* 12, 814, 1976.

215. **S. V. Nemilov** "Interrelation between the Sound Spreading Speed, Mass and Energy of Chemical Reaction" *Dokl. Akad. Nauk SSSR* 181, 1427, 1968.

216. **A. Feltz, B. Voigt, L. Senf, G. Dresler** "Uber Neue Infrarotdurchlossige Optische Glaser" Wiss. Z. Friedrich Schiller Univ. Jena Math. Naturwiss. Reiche 28, 327, 1979.

217. **J. M. Lloyd** "Thermal Imaging Systems" Plenum Press, New York, 1975.

218. **A. Feltz, W. Burckardt, B. Voigt, D. Linke** "Optical Glasses for IR-Transmittance" *J. Non-Cryst. Solids* 129, 31, 1991.

219. "The Buyers Guide, Laser Focus World" 1994, 614.

220. Optical Details Made of Oxygen-free Glass. Standard Technological Process of Mechanical Treatment. Branch Standard OST 3–5792–85, 1985.

Author Bibliography

1. **L. G. Aijo, V. F. Kokorina** "Optical Glasses, Transparent in Infrared Spectrum Range to l = 11–15 mm. I. Method of Laboratory Production of Sulpho-Selenide Optical Glasses" *Opt. Mekh. Promst.* 4, 39, 1961.

2. **V. F. Kokorina** "On Crystalline Formations in Oxygen-Free Sulfo-Selenide Glasses" *Zh. Prikl. Khym.* 38, 1472, 1965.

3. **L. G. Aijo, V. F. Kokorina** "Glass-Formation Principles and Some Properties of Sulfo-Selenide Glasses in Elementary Systems" *OMP* 6, 48, 1961.

4. **L. G. Aijo, V. F. Kokorina** "Glasses, Containing Chalcogenides of IV and V Groups of Periodic System" *OMP* 5, 32, 1963.

5. **A. M. Yefimov, V. A. Khariyuzov** "Dielectrical Properties and Structure of Chalcogenide Glasses in the Systems Arsenic-Selenium and Germanium-Selenium" *Stekloobraznoije Sostoyanije Isd. Nauka,* Moscow-Leningrad, 1971, 370.

6. **Ye. A. Yegorova (Kislitskaya), V. F. Kokorina** "Glass-Formation and Physiko-Chemical Properties of Glasses in the System As–P–Se" *OMP* 1, 33, 1963.

7. **L. G. Aijo, V. F. Kokorina** "Glass-Formation and Properties of Glasses in the System As–Ge–Se" *OMP* 2, 36, 1963.

8. **V. A. Khariyuzov, V. F. Kokorina, M. V. Proskuryakov, V. K. Borina, Ye. A. Kislitskaya** "Dielectrical Properties and Structure of Glasses in the System As–Ge–Se" Collection "Svoistva i Razrabotka Novykh Opticheskikh Styokol" Isd. Mashinostroyenije, Leningrad, 1977, 127.

9. **Ye. A. Yegorova (Kislitskaya), V. F. Kokorina** "Glass-Formation and Physico-Chemical Properties of Glasses in the System Sb–Ge–Se" *Zh. Prikl. Khym.* 41, 1200, 1968.

10. **V. F. Kokorina, A. M. Tyutinov** "Secondary Electron Emission of Quenched Chalcogenide Glasses" *Phys. Khym. Stekla* 4, 739, 1978.

11. **A. M. Yefimov, V. F. Kokorina** "Some Properties and Structure of Oxygen-Free Glasses, Containing Heavy Elements of the IV Group of the Periodic System" Stekloobraznoie Sostoyanije, Isd. Nauka, Moscow-Leningrad, 1965, 177.

12. **A. M. Yefimov, V. F. Kokorina** "Glass-Formation and Properties of Glasses in the System As–Ge–Sn–Se" *OMP* 12, 37, 1965.

13. **A. M. Yefimov, V. F. Kokorina** "Glass-Formation and Properties of Glasses in the System As–Ge–Pb–Se" *OMP* 10, 43, 1969.

14. **A. M. Yefimov** "Glass-Formation and Some Properties of Glasses in the System Ge–Sn–Pb–As–Se" Collection "Opticheskoie Steklo" Isd. Mashinostroienije, Leningrad, 1972, 100.

15. **Ye. A. Kislitskaya, V. F. Kokorina** "The Influence of Germanium Replacement with Tin on Glass-Formation and Physico-Chemical Properties of Glasses in the System Sb–Ge–Se" *Zh. Prikl. Khym.* 44, 646, 1971.

16. **L. G. Aijo, V. F. Kokorina** "Glass-Formation and Properties of Glasses in the System Ge–Se–Te" Collection "Issledovanije Stekloobrazovanija i Sintez Novykh Styokol" Isd. VIINITI, Minsk-Moscow, 1971, 19.

17. **V. F. Kokorina, V. V. Melnikov** "Synthesis, Properties and Application of Oxygen-Free Halogencontaining Glasses, Transparent in IR-Spectrum Range" *Collection Zharoprochnyie Neorganicheskije Materialy,* Isd. NIITS, 1977, 283.

18. **A. M. Yefimov, V. F. Kokorina** "On Correlation of the Near Order Structure in Glass and Crystal" *Stekloobraznoie Sostoyanije, Isd. Nauka,* Leningrad, 1971, 92.

19. **V. F. Kokorina** "Some Problems of Glass Structure According to the Evidence from Oxygen-Free Glasses Investigation" *Stekloobraznoie Sostoyanije, Isd. Nauka,* Moscow-Leningrad, 1965, 174.

20. **V. F. Kokorina** "Influence of Chemical Bond on the Glass-Formation and Properties of Glasses" *Stekloobraznoie Sostoyanije, Isd. Nauka,* Leningrad, 1971, 87.

21. **L. G. Aijo, A. M. Yefimov, V. F. Kokorina** "Refractive Index of Chalcogenide Glasses with the Wide Range of Compositions" *J. Non-Cryst. Solids* 27, 299, 1978.

22. **L. G. Aijo, G. S. Gorbunova, V. F. Kokorina, V. V. Melnikov** "Thermal Increment of Refractive Index of Oxygen-Free Glasses" *OMP* 11, 36, 1976.

23. **V. F. Kokorina, L. G. Aijo, Ye. A. Kislitzkaya, V. V. Melnikov** "New Investigations of Glass-Formation and Properties of Oxygen-Free Semiconductor Glasses" Struktura i Svoistva Nekristallicheskikh Poluprovodnikov. Trudy VI Mezhdunarodnoi Konferencii po Amorfnym i Zhidkim Poluprovodnikam Isd. Nauka, Leningrad, 1976, 39.

24. **A. V. Khomenko (Belykh), G. M. Orlova, N. S. Martynova** "Calculated and Experimental Determination of the Eutectic Point in the System Sb_2Se_3–$GeSe$–$GeSe_2$" Zh. Neorg. Khim. 27, 485, 1982.

25. **Ye. A. Kislitskaya, V. B. Nosov, V. F. Kokorina** "Optical Absorption in Oxygen-Free Glasses Based on Arsenic, Germanium and Selenium" *Phys. Khim. Stekla* 3, 624, 1977.

26. **Ye. A. Kislitskaya, V. F. Kokorina, A. V. Shatilov** "Influence of the Structure of Chalcogenide Glasses on the Character and Edge Powers of Their Surfaces' Destruction" *Phys. Khim. Stekla* 9, 177, 1983.

27. **Z. I. Kanchijev, V. F. Kokorina** "Impurities' Removal out of the Vitreous Arsenic Sulfide" *OMP* 9, 45, 1975.

28. **V. F. Kokorina, Z. I. Kanchijev, A. V. Belykh** "Multicomponent Chalcogenide Glasses for IR-Fiber" Collection Physico-Opticheskije Svoistva Materialov Dlya Volokonnykh Svetovodov. Isd. NPO Stekloplastik, Moscow, 1988, 70.

29. **Ye. I. Melnikova, V. V. Melnikov, L. V. Sergeev** "Thermoplastic Gluing Fusion TKS1, Transparent in IR-Spectrum Range" *OMP* 3, 68, 1973.

30. **V. V. Melnikov, V. F. Kokorina** "Elaboration and Investigation of IR-Transparent Chalcogenide Glasses for Spectroscopy NPVO" Collection "Analiticheskoie Priborostroyenije" Tbyilisi, 1980, 138.

31. **V. F. Kokorina, V. V. Melnikov, Ye. M. Milyukov** "Oxygen-Free Glass-Filter for Infrared Spectrum Range" Collection "Zharoprochnye Heorganicheskije Materialy" Isd. NIITS, 1977, 280.

32. **I. M. Adrianova, L. G. Aijo, L. N. Asnis, Ye. A. Kislitskaya, A. V. Moskalenko** "Acoustic-Optical Properties of Glasses in the System As–Se" *Akust. Zh."* 21, 822, 1975.

33. **I. I. Adrianova, L. G. Aijo, L. N. Asnis, Ye. A. Kislitskaya, V. F. Kokorina** "Acoustic-Optical Properties of Glasses in the Systems As–Ge–Se and As–Ge–Se–Sb" *Akust. Zh.* 22, 449, 1976.

34. **I. M. Adrianova, L. N. Asnis, V. V. Melnikov, A. V. Petrova** "Investigation of Acoustic-Optical Properties of Glasses in the System As–S–I–Sb" *Opt. Spektrosk.* 37, 782, 1974.

35. **L. G. Aijo, V. A. Arephiev, V. F. Kokorina, G. V. Shkal'kova** "The Influence of Cooling Speed of Oxygen-Free Chalcogenide Glasses in the Region of Critical Annealing on the Optical Constants" *OMP* 10, 29, 1979.

36. **Ye. A. Kislitskaya, V. F. Kokorina** "Oxygen-Free Infrared Optical Glasses" Catalog "Tzvetnoye Opticheskoie Steklo i Osobyje Styokla" Isd. Dom Optiki, Moscow, 1990, 177.

37. **L. G. Aijo, Ye. A. Kislitskaya, V. F. Kokorina** "Optical Oxygen-Free Glass" Branch Standard, OST3–3441–83 1983.

Index